GEOTECHNICAL SPECIAL PUBLICATIONS

1) TERZAGHI LECTURES
2) GEOTECHNICAL ASPECTS OF STIFF AND HARD CLAYS
3) LANDSLIDE DAMS: PROCESSES, RISK, AND MITIGATION
4) TIEBACKS FOR BULKHEADS
5) SETTLEMENT OF SHALLOW FOUNDATION ON COHESIONLESS SOILS: DESIGN AND PERFORMANCE
6) USE OF IN SITU TESTS IN GEOTECHNICAL ENGINEERING
7) TIMBER BULKHEADS
8) FOUNDATIONS FOR TRANSMISSION LINE TOWERS
9) FOUNDATIONS AND EXCAVATIONS IN DECOMPOSED ROCK OF THE PIEDMONT PROVINCE
10) ENGINEERING ASPECTS OF SOIL EROSION, DISPERSIVE CLAYS AND LOESS
11) DYNAMIC RESPONSE OF PILE FOUNDATIONS— EXPERIMENT, ANALYSIS AND OBSERVATION
12) SOIL IMPROVEMENT - A TEN YEAR UPDATE
13) GEOTECHNICAL PRACTICE FOR SOLID WASTE DISPOSAL '87
14) GEOTECHNICAL ASPECTS OF KARST TERRAINS
15) MEASURED PERFORMANCE SHALLOW FOUNDATIONS
16) SPECIAL TOPICS IN FOUNDATIONS
17) SOIL PROPERTIES EVALUATION FROM CENTRIFUGAL MODELS
18) GEOSYNTHETICS FOR SOIL IMPROVEMENT
19) MINE INDUCED SUBSIDENCE: EFFECTS ON ENGINEERED STRUCTURES
20) EARTHQUAKE ENGINEERING & SOIL DYNAMICS (II)
21) HYDRAULIC FILL STRUCTURES
22) FOUNDATION ENGINEERING
23) PREDICTED AND OBSERVED AXIAL BEHAVIOR OF PILES
24) RESILIENT MODULI OF SOILS: LABORATORY CONDITIONS
25) DESIGN AND PERFORMANCE OF EARTH RETAINING STRUCTURES
26) WASTE CONTAINMENT SYSTEMS: CONSTRUCTION, REGULATION, AND PERFORMANCE
27) GEOTECHNICAL ENGINEERING CONGRESS
28) DETECTION OF AND CONSTRUCTION AT THE SOIL/ROCK INTERFACE
29) RECENT ADVANCES IN INSTRUMENTATION, DATA ACQUISITION AND TESTING IN SOIL DYNAMICS
30) GROUTING, SOIL IMPROVEMENT AND GEOSYNTHETICS
31) STABILITY AND PERFORMANCE OF SLOPES AND EMBANKMENTS II (A 25-YEAR PERSPECTIVE)
32) EMBANKMENT DAMS-JAMES L. SHERARD CONTRIBUTIONS
33) EXCAVATION AND SUPPORT FOR THE URBAN INFRASTRUCTURE
34) PILES UNDER DYNAMIC LOADS
35) GEOTECHNICAL PRACTICE IN DAM REHABILITATION
36) FLY ASH FOR SOIL IMPROVEMENT
37) ADVANCES IN SITE CHARACTERIZATION: DATA ACQUISITION, DATA MANAGEMENT AND DATA INTERPRETATION

Innovative Design and Construction for Foundations and Substructures Subject to Freezing and Frost

Proceedings of a session sponsored by
The Geo-Institute of the American Society of Civil Engineers
in conjunction with the
ASCE National Convention, Minneapolis, Minnesota,
October 5-8, 1997

Edited by Chia K. Tan

Geotechnical Special Publication No. 73

Published by the
ASCE American Society
of Civil Engineers
1801 Alexander Bell Drive
Reston, VA 20191-4400

Abstract:
This proceedings, *Innovative Design and Construction for Foundations and Substructures Subject to Freezing and Frost*, consists of papers presented at a session sponsored by The Geo-Institute of the ASCE in conjunction with the National Convention held in Minneapolis, Minnesota, October 5-8, 1997. The publication's primary objective is to share information gained from recent research findings and experience involving foundations and substructures subjected to freezing and permafrost conditions. Since the traditional approach of embedding foundations or substructures below the historical depth of the frost penetration is not the most cost-effective solution, new design and construction methods are being developed. Three out of the four papers present research concerned with these new methods that emphasize modifying one or more of the critical elements needed for the frost action to occur. The fourth paper describes the status of the proposed ASCE standard on frost protected shallow foundations.

Library of Congress Cataloging-in-Publication Data

Innovative design and construction for foundations and substructures subject to freezing and frost : proceedings of a session sponsored by the Geo-Institute of the American Society of Civil Engineers in conjunction with the ASCE National Convention, Minneapolis, Minnesota, October 5-8, 1997 / edited by Chia K. Tan.
p.   cm. -- (Geotechnical special publication ; no. 73)
Includes index.
ISBN 0-7844-0291-4
1. Foundations--Cold weather conditions--Congresses. 2. Basements--Cold weather conditions--Congresses. 3. Foundations--Cold regions--Design and construction--Congresses. 4. Basements--Cold regions--Design and construction--Congresses. I. Tan, Chia K. II. American Society of Civil Engineers. Geo-Institute. III. ASCE National Convention (1997 : Minneapolis, Minn.) IV. Series.
TH2101.I53 1997   97-36204
624.1'5--dc21   CIP

A significant design constraint for foundations and substructures which are constructed in temperate zones of the world, which include the northern half of the United States and all of Canada, is the provision for protecting these foundations and substructures from the harmful effects of the frost action in the soils. The traditional approach has been to embed the foundations or substructures below the historical depth of the frost penetration. This approach is not necessarily the most cost-effective solution, and it may totally disregard the effective utilization of precious resources.

In response to these concerns, new design and construction methods are being developed. The design and construction methods approach the problem by trying to modify one or more of the critical elements needed for the frost action to occur: the presence of water, freezing conditions, and frost susceptible soil conditions. Significant progress has been made over the last two decades. As a result, a new ASCE design standard is presently being developed for design and construction of frost protected shallow foundations.

This special technical publication constitutes the proceedings of a technical session which was held at the American Society of Civil Engineers (ASCE) National Convention held in Minneapolis, Minnesota, in October 1997. The technical session entitled " Innovative Design and Construction for Foundations and Substructures Subjected to Freezing Conditions" was sponsored by the Shallow Foundation Committee and Cold Region Engineering Technical Committee of ASCE. The primary objective of the technical session was to share information gained from recent research findings and experience involving foundations and substructures subjected to freezing and permafrost conditions. The status of the proposed ASCE standard on frost protected shallow foundations was also described in one of the papers.

It is the current practice of the Geotechnical Engineering Division that each paper published in a Geotechnical Special Publication (GSP) be reviewed for its content and quality. These GSPs are intended to reinforce the programs presented at convention sessions or specialty conferences and to contain papers that are timely and may be controversial to some extent. The time available for reviews is generally not as long as that given to papers submitted to the *Journal of Geotechnical and Geoenvironmental Engineering* due to the need to have the GSP available at the time of the Convention. However, in accordance with the ASCE policy, the papers for this GSP received at least two positive peer reviews and the authors were given the opportunity to modify their papers based on reviewers' suggestions prior to final submittal of the papers. All papers included in this GSP are eligible for discussion in the *ASCE Journal of Geotechnical and Geoenvironmental Engineering* and are eligible for ASCE awards.

The editor of this GSP expresses his gratitude to all of the reviewers and authors who contributed a significant amount of time and effort to make the publication of this proceedings possible. Special thanks are due to Kent Wray, Chairman of the Shallow Foundations Committee, for his help and suggestions in organizing this session. Thanks are also extended to Shiela Menaker and Barb Vanderheiden for their patience and help in handling and organizing the papers in this GSP.

Chia K. Tan
STS Consultants, Ltd.

CONTENTS

Significant Soil Properties in the Thermal and
Structural Design of Building Foundations

Louis F. Goldberg Ph.D[1]

Abstract

A specification of significant soil properties necessary for the thermal and structural design of building foundations has been developed based on many years of experimental testing carried out at the University of Minnesota's Foundation Test Facility. A brief review of some experimental results describing the thermal transport through frost-protected shallow foundations is given to illustrate the context in which the soil properties specification was developed. This is followed by two case studies illustrating the application of the specification to the design of a frost-protected foundation, and, to the geotechnical investigation of a rock glacier building site for the purpose of providing baseline foundation design data.

Introduction

Since commencing operations in 1988, the University of Minnesota's Foundation Test Facility (FTF) located in Rosemount, MN, has yielded a wealth of experimental data that have been used to improve and develop techniques for the thermal and structural design of conventional and innovative building foundation systems. In order to generalize the data from the specific conditions pertaining at the FTF to any foundation system, three-dimensional computer simulation techniques have been tested against the experimental data base. In order to obtain satisfactory agreement between the simulated and experimental data, the soil properties need to be appropriately specified. In this regard, a specification of the needed soil properties has been developed and applied in commercial building foundation design practice.

An overview of the FTF and an example of some thermal performance

----

[1]Principal, Lofrango Engineering, and Senior Research Associate, Univ. of Minnesota, 5214, 10th Avenue South, Minneapolis, MN 55417

1

results are presented first in order to demonstrate the experimental context in which the soil properties specification was developed.   This is followed by two case studies illustrating the use of the soil properties specification, firstly to the thermal design of a foundation system, and, secondly, to a geotechnical investigation of a particular site.

The Foundation Test Facility

The FTF was established in 1988 at the University of Minnesota's Rosemount Research Center.   The primary purpose of the facility is to demonstrate objectively and unambiguously that foundation insulation produces significant energy savings, particularly in a cold Minnesota climate. In order to generalize the results obtained at the FTF to building foundations throughout Minnesota, it is necessary to develop appropriate techniques for calculating the earth-contact building envelope heat transfer for all the soil and climatic conditions and foundation configurations found throughout the state. This has been accomplished using energy and moisture transport computer simulation codes.   Hence the FTF also has been used to generate sufficient three-dimensional experimental data to test the physical viability and accuracy of these simulation codes.

The basic layout of the FTF is depicted in figure 1. The facility comprises six test modules, each having a square footprint of approximately 37.2 m^2 (400 ft^2) in area. The modules are positioned on a staggered grid with an inter-module separation greater than 12.2 m (40 ft). This arrangement ensures similarity of microclimates around each module for the prevailing northwesterly wind direction while providing earth-contact heat transfer separation between the modules. Four of the modules are full basements while the remaining two modules have a shallow slab-on-grade / stem wall configuration. An instrumentation trailer to the west of the modules completes the installation.

The experimental protocol is based in essence on a three-dimensional evaluation of each foundation system being tested.   In order to allow comparison of results between heating seasons, each foundation type includes 'test' and 'reference' modules (figure 1). Thus, the results obtained from the test modules are normalized against those of the reference modules allowing comparison of the normalized results of different foundation systems collected over different heating seasons. The reference modules remain unchanged and are constructed to be as simple a physical realization of a foundation type as possible.

The overall design of the shallow foundation modules is shown in figure 2. The soil environment surrounding these modules is engineered to be the same. The native soil was excavated to a depth of 1.1 m (42 in.) for both the test and reference modules so that the base of the excavation is in a sand or

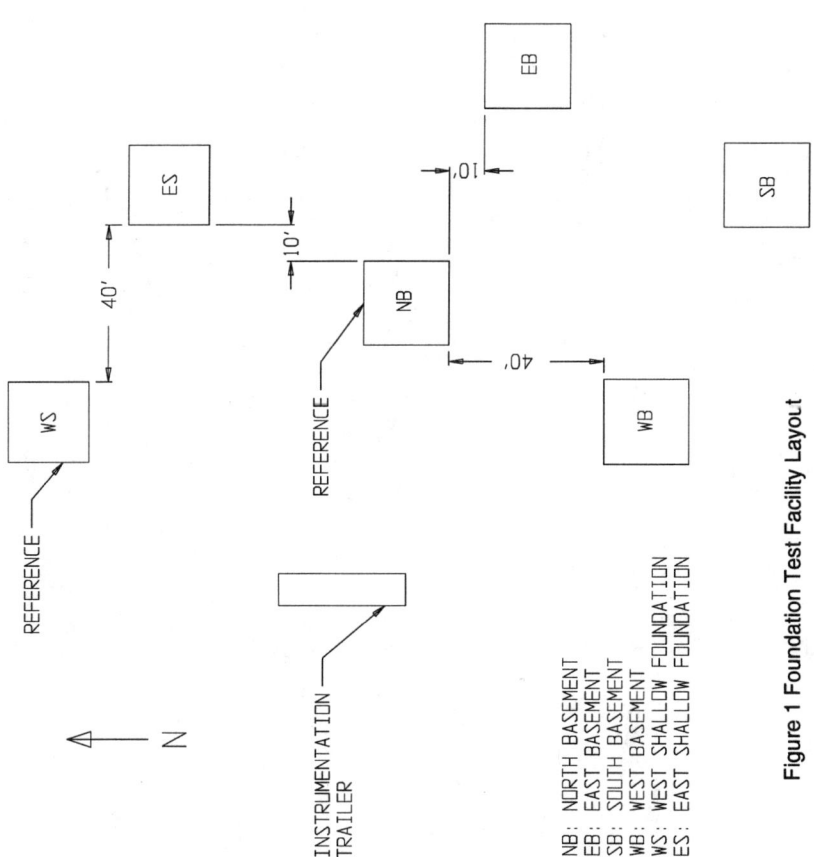

Figure 1 Foundation Test Facility Layout

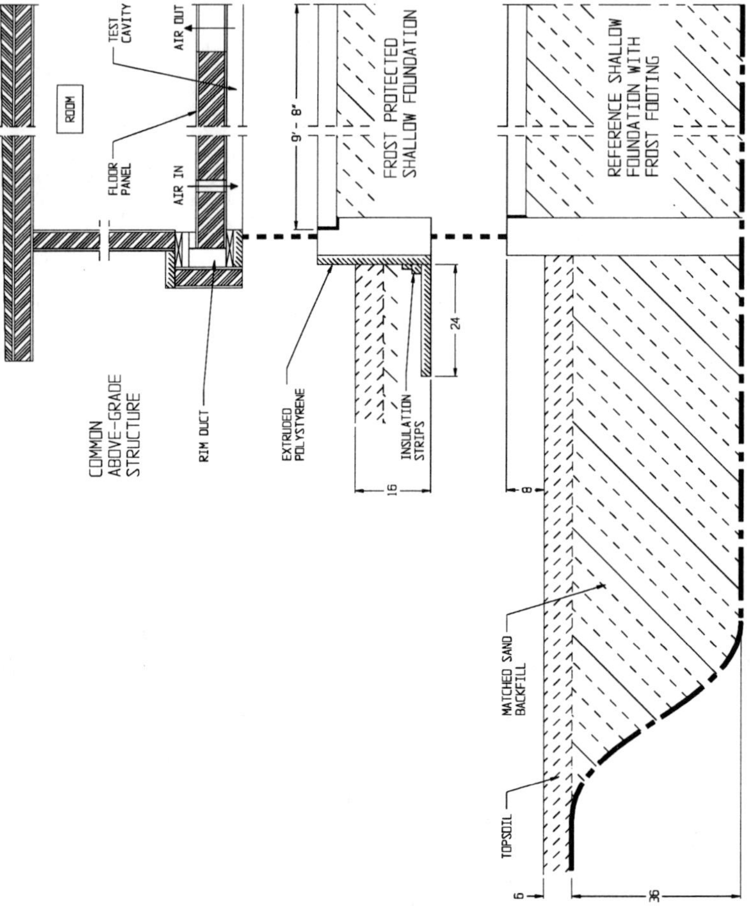

Figure 2 Shallow Foundation Module Design

sand/gravel stratum. The excavations are backfilled with a uniform sand, and a 152 mm (6-in.) layer of native agricultural loam above the backfill allows the growth of indigenous vegetation.

The reference module foundation is built in accordance with the Minnesota amendments to the 1988 Uniform Building Code (*MN Dept. Of Admin., 1990*) and consists of a nominally 203 mm (8 in.) thick stem wall with its base 1.1 m (42 in.) below grade. In order to maintain the requirement of class simplicity, no steel reinforcement is used and the walls bear directly on a compacted sand base without a spread footing. The shallow foundation module design was based on the principals elucidated in the Comite Europeen de Normalisation (CEN) proposed standard N185 (*Comite Europeen, 1992*). These standards have found their way via the HUD design guide (*NAHB, 1994*) into the CABO standard in the United States. In essence, this standard stipulates the quantity and location of stem wall insulation necessary to direct heat leaking from a continuously heated building to the footing region to maintain the sub-footing temperatures above freezing, even during 100-year return adverse ambient temperature conditions. The above-slab structures for both shallow foundation modules are identical and are designed to allow the net heat transfer to the slab to be isolated and measured experimentally. Details of the above-slab structure may be found in *Goldberg, Langenfeld et al, 1994.*

Figure 3 shows the thermal performance of the shallow foundation modules for the 1991/92 and 1992/93 heating seasons. The smallness of the error bars indicates the good repeatability of the thermal performance experimental protocol. The left hand bars show that in terms of whole envelope energy transport, the east (module ES) frost-protected shallow foundation module lost 14 % less energy than the conventional frost-footing case (module WS). However, isolating the slab thermal transport, reveals that this transport amounted to 19 % of the total envelope transport for module WS and 13 % of the total for module ES. Thus a comparison of the slab thermal transport alone in the right hand bars reveals that the frost protected shallow foundation system reduced the slab heat loss by 41 % on a normalized basis compared with the conventional (uninsulated) frost footing foundation.

In applying thermal transport simulation codes to the FTF modules with the intent of replicating thermal performance results of the type discussed above, the following soil properties have been found to be satisfactory for obtaining reasonable agreement between measured and simulated data:

- dry density
- dry heat capacity
- porosity
- dry thermal conductivity
- saturation ratio

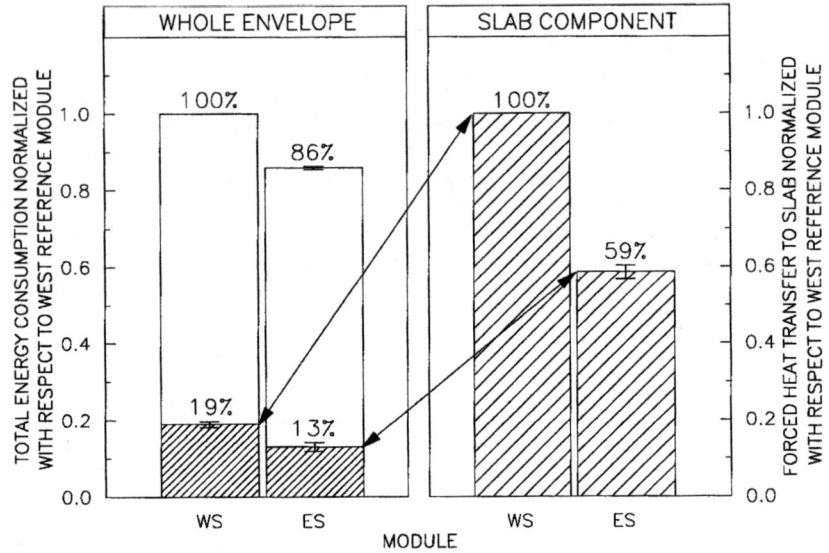

Figure 3 Thermal Performance of Shallow Foundation Modules
(1991/92 and 1992/93 heating seasons)

- strain configuration (orientation of ice lens minor axis)

These properties are expressed as functions of spatial location and, if transient, as a function of time as well. Thus, in general, each soil or rock type present in the soil domain will have a separate transient soil property specification. The mechanisms by which the soil properties are included in a given simulation code depend on the capabilities of the code. Some codes allow the baseline dry property data to be converted into actual values as a function of moisture content, temperature and stress. Other codes provide a subset of this generality, while, in most cases, only a limited set of fixed properties can be specified (usually, heat capacity, thermal conductivity and density). In the latter case, the dry baseline properties need to be combined in a pre-processing stage to yield effective wet values. Sometimes the transient variation of these values can be accommodated via a schedule, while in other instances, temporally averaged values must be used.

Case Study 1: Design of a frost-protected foundation for an unheated building

A lake home on the shore of Sand Lake (located about 50 miles E-NE of Bemidji, MN) required a foundation system that would permit the home to remain unheated during the unoccupied winter season. The site is located in the 3000-3500 FFDD (Fahrenheit Freezing Degree Day) 100-year return air freezing Index (AFI) zone so that the design AFI selected was 3500 FFDD. The design weather data set was derived by statistically modifying the Fargo, MN TMY weather data set to yield a 3500 FFDD AFI using a mean depression technique.

The home has a two-storey layout consisting of a main floor and a loft upper story resting upon a walk-out full basement with a rectangular 8.5 m by 11.0 m (28 ft by 36 ft) floor plan. The basement has a fully-bermed west elevation, with the berms tapering along the south and north sides to a slab-on-grade configuration along the east-facing, lakeside elevation. The southern half of the basement is devoted to a single-car garage, the remainder to a bathroom, sauna, utilities and storage. Topographically, the site is on a steeply sloping bank, so that the basement floor is much higher than the median lake surface water level.

Soil samples collected around the building perimeter indicated a fairly uniform sandy loam, approximating P4711 Dakota sandy loam (*Kersten, 1948*). Owing to the small range of variation encountered plus the intention of using excavated soil as a backfill, the "worst thermal case" soil properties shown in table 1 could be used for the entire soil domain. In terms of the relatively high thermal conductivity of the soil and the small variations encountered within the soil sample set, this choice allowed a reduction in the number of discrete soil descriptions required for the simulation while retaining a conservative design posture, so reducing design costs.

The worst case moisture content profile measured is given in terms of the saturation ratio in table 2. The soil properties described in tables 1 and 2 are sufficient to define the soil domain for the particular foundation heat transfer simulation code[1] used in this study without any pre-processing.

Table 1 Design soil properties

Dry density	1689 kg/m³
Dry heat capacity	837 J/kg.K
Porosity	.377
Dry thermal conductivity	.264 W/m.K
Strain configuration	anisotropic

Of particular note in table 1, is the specification of the dry thermal conductivity as the parameter used to define the soil conductivity. The definition allows the code to compute spatially and temporally varying conductivities as a function of temperature, saturation ratio and ice content using Johansen's method (*Farouki, 1986*).

Table 2 Design soil moisture contents

Below grade depth (mm)	Saturation ratio	Below grade depth (mm)	Saturation ratio
152	.65	1295	.48
381	.40	1448	.53
533	.38	1600	.70
686	.41	1753	.59
838	.51	1905	.58
991	.39	2057	.50
1143	.47	7620	1.00

Stipulating an anisotropic strain configuration allows the code to displace the soil domain preferentially along the minor axes of ice lenses as they form while allowing for the isotropic strain produced by soil dilation. The soil moisture content is expressed in terms of the void volume based saturation ratio rather than the conventional mass based moisture content as this is computationally more convenient for describing transient changes in soil moisture produced by

— — — — — — — — — — — —

[1]Building Foundation Energy Transport Simulation (BUFETS) rev. B.4, a proprietary, non-commercial code.

frost heave (ice lens formation) and soil dilation. The soil saturation ratio described in table 2 follows that measured down to the 2057 mm (6.75 ft) level after which it increases to unity (soil saturation) at the 7620 mm (25 ft) level.

The final design developed that meets the primary requirement of allowing the house to remain unheated through a 100 year return winter and survive without frost-heave or soil-dilation induced structural damage is shown in figures 4 and 5. Figure 4 shows a cross-section through the fully bermed wall on the west side of the basement. The layout developed uses formed-in-place poured concrete walls with RSI-1.8 (R-10) extruded polystyrene insulation as the permanent forms. A 102 mm (4 in.) slab is poured above a 76 mm (3 in.) sand water absorption layer and RSI-1.8 (R-10) extruded polystyrene insulation. A 127 mm (5 in.) gravel drainage layer supports a conventional spread footing while the drainage layer is fitted with a drain pipe routed to a sump. This drainage system provides better soil moisture control and increases the effective non-frost-susceptible foundation depth.

A cross-section through the stem-wall (walk-out) portion of the basement on the east side is given in figure 5. The layout is similar to that of figure 4 with the exception of RSI-2.6 (R-15) thermal insulation extending beneath the stem wall to a wing width of 762 mm (30 in.). A simplified cut-away view of the foundation system around the southeast corner is shown in figure 6. The figure shows the maximum frost front penetration at the outer edge of the footing along the south and east walls. Of particular note is the manner in which the foundation system on the partially bermed south wall undergoes a transition at the point where the berm recedes from the top of the wall. The frost-protection system is designed to take advantage of the installed drainage whereby water is drained away from the rear portion of the wall to the drain tile beneath the front (wing-insulated) portion of the wall. This ensures that the saturation ratio of the soil beneath the rear portion of the wall is less than the critical saturation ratio (about .92), so that even if the ground freezes (as shown by the frost line location beneath the footings of the rear portion of the wall), there is no danger of frost heave or soil dilation taking place.

This concept involves what has been termed a "stage II" frost-protected foundation design strategy (as opposed to the purely "stage I" CABO approach described above), in which the governing principle is that the saturation ratio of the soils in all frozen regions beneath the footings shall always be less than the critical saturation ratio. It also should be noted that the foundation meets the applicable Minnesota amendments to the 1988 Uniform Building Code (*MN Dept. of Admin., 1990*) whereby the base of the foundation is at least 1.5 m (60 in.) below grade at all points around the building perimeter. However, as shown in figure 6, in this case, meeting the code is, by itself, insufficient to prevent frost from penetrating beneath the footing at certain points around the foundation perimeter. Finally, an exploded view of the southeast corner showing the point

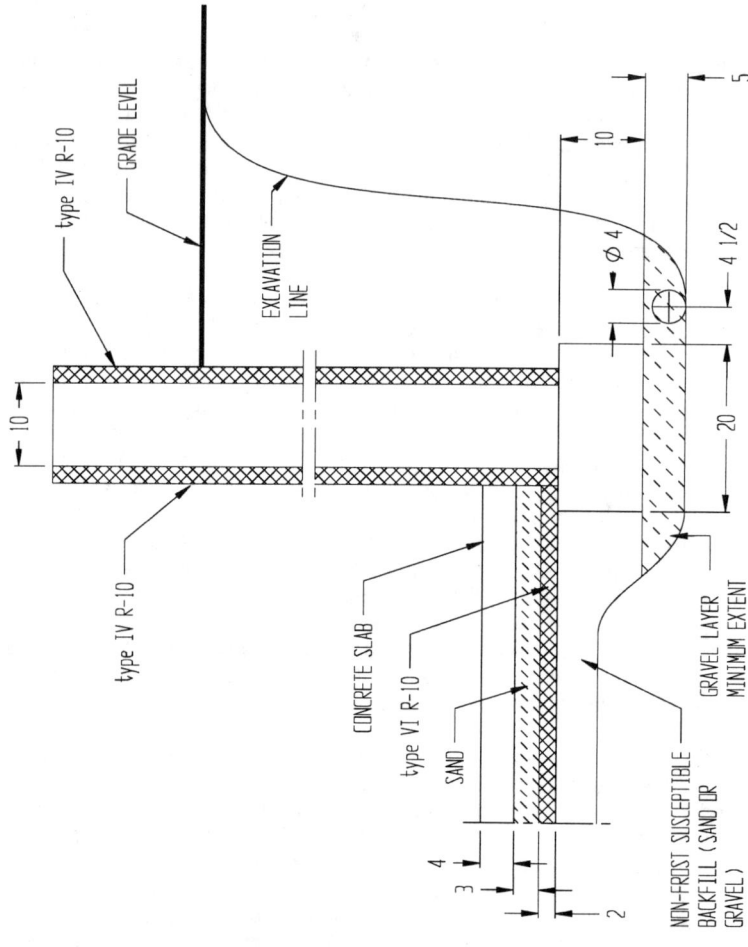

Figure 4 Cross Section Through Rear Foundation Wall

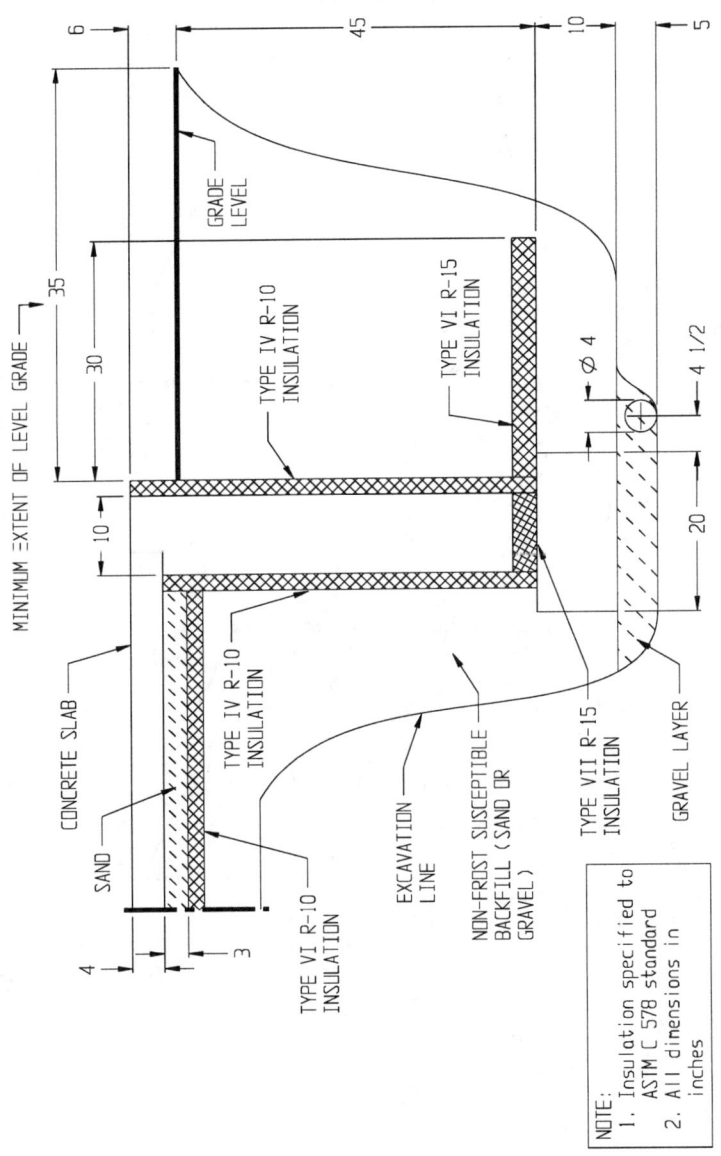

Figure 5 Unheated Walk-out Basement Stem Wall Schematic

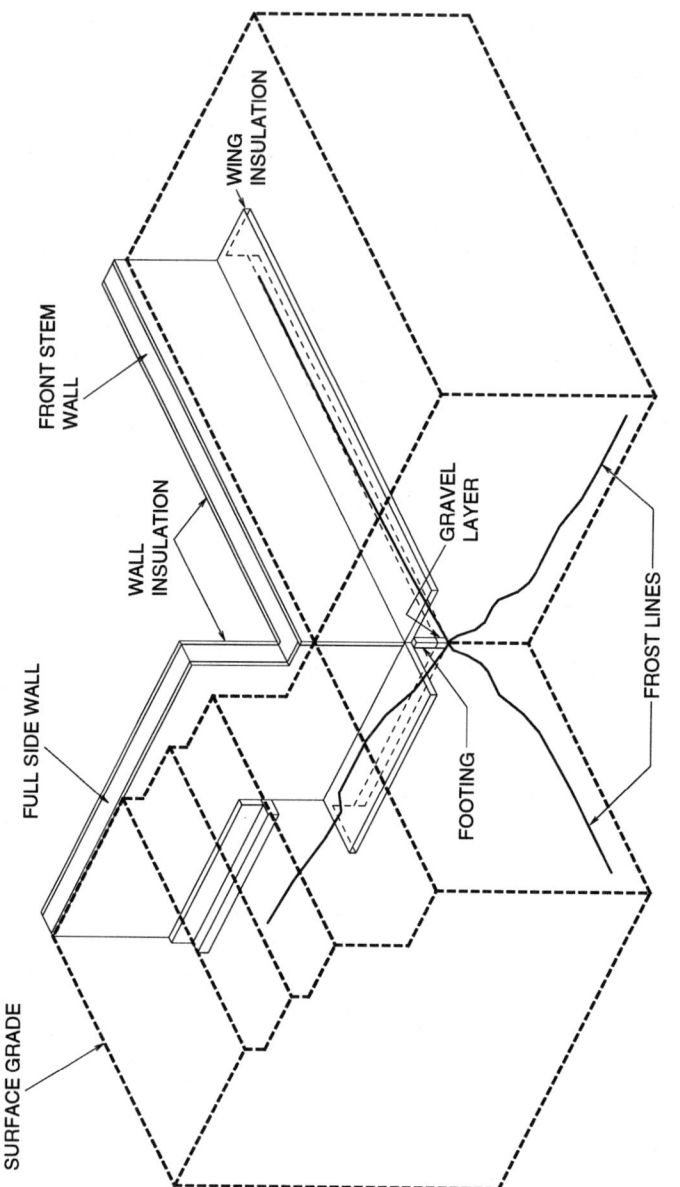

Figure 6 Unheated Walk-out Basement Foundation Maximum Frost Penetration (at 1h00 on 4/7)

of maximum frost penetration is given in figure 7. The design was tuned to yield a zero frost front clearance right at the corner in order to provide the homeowners with the minimum depth of foundation that would provide frost protection. At their discretion, the homeowners thus were able to increase the depth of the gravel layer and/or the thickness of the wing insulation depending on their budget and the additional margin of safety they felt comfortable with.

Case study 2: Geotechnical investigation of a building site atop a rock glacier

It is proposed to construct a new hospitality building on the summit of Pikes Peak in Colorado Springs, CO to replace the existing building which is experiencing gradual structural failure as a result of thermally produced foundation subsidence. The top of Pikes Peak has a unique geology consisting of the following layers in a bottom-down sequence: fill; boulder rubble with gravel and sand; frozen silty sand; a gravel, rock and sand impregnated ice sheet; and, porphyritic granite. Thus the foundation design required is complex since it needs to maintain the thermal integrity of the ice sheet while simultaneously providing adequate bearing strength at a reasonable cost.

In support of the design effort, a preliminary geotechnical survey of the site was carried out in which a series of four test pits were excavated around the proposed building perimeter. The raw field data gathered are given in table 3.

In transforming the raw data into the defined soil properties specification it is necessary to establish a sample soil volume basis. In general, two soil volumes are suitable as the basis, namely the in-situ volume of the soil sample and the water-saturated laboratory test volume. For typical soils that experience annual freeze/thaw cycles, these two basis volumes are quantitatively similar. However, in situations where the soil has remained undisturbed for a prolonged period, and has not been subject to freeze thaw cycles, then the amount of soil consolidation can be significant yielding an in-situ volume less than the saturated volume. In contrast, an in-situ volume larger than the saturated test volume is a clear indication of an occurrence of frost heave. But such a volume difference that is contextually too large for a homogeneous sample without large voids is not physically reasonable since the supervening soil mass would compress any large voids over time. Such a large volume difference is thus usually an indication of a systematic error. The Pike's Peak situation is typical of that in which the use of the in-situ volume calculation basis is appropriate. However, in order to enable the physical validity of the results to be assessed, all the calculations of the soil properties were performed on both the in-situ and saturated test volume bases.

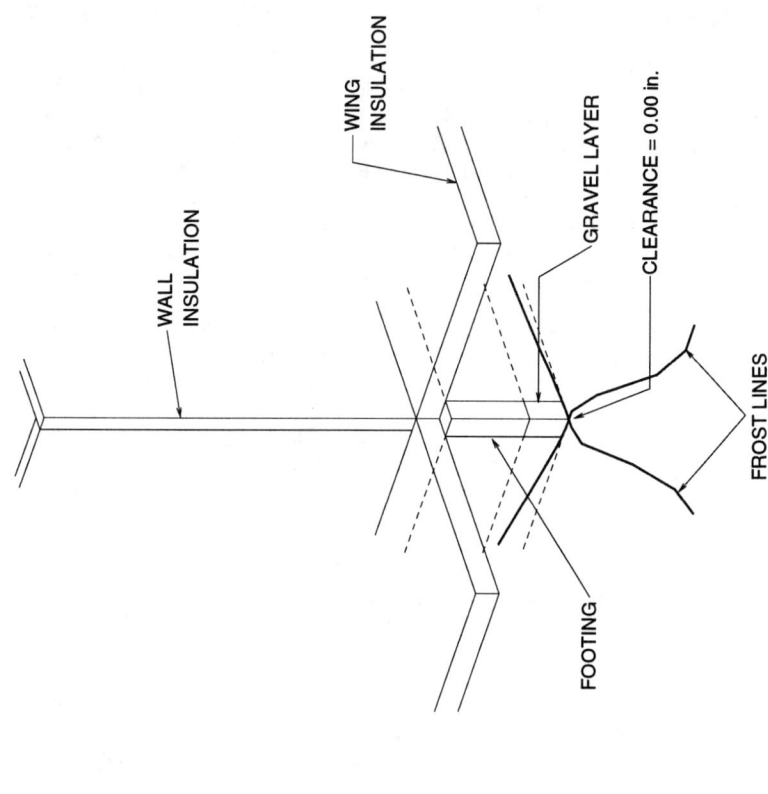

Figure 7 Unheated Walk-out Basement Foundation Maximum Frost Penetration Detail (at 1h00 on 4/7)

Table 3 Pike's Peak geotechnical survey raw data

Sample Location	Elevation (ft)	Temp. (°F)	Wet Mass (g)	Dry Mass (g)	Saturated Mass (g)	In-Situ Volume (ft³)	Specific Gravity
TP 1 - surface	97.73						
TP 1 - 1	95.50	31.5	1974.7	1852.9	2253.4	0.0335	2.57
TP 1 - 2	94.88	31.0	1268.5	1172.1	1471.8	0.0252	2.6
TP 1 - 3	93.41	30.5	847.1	768.8	983.6	0.0136	2.59
TP 1 - 4	93.50	31.7	2253.5	2017.4	2484.8	0.0398	2.53
TP 1 - 5	92.98	30.6	979.7	873.4	1107.2	0.0206	2.59
TP 2 - surface	96.73						
TP 2 - 1	90.86	31.2	1207.8	1127.7	1325.3	0.0209	2.54
TP 2 - 2	90.13	31.1	914.7	805.0	1035.1	0.0264	2.59
TP 2 - 3	85.91	30.7					
TP 3 - surface	96.68						
TP 3 - 1	90.78	31.8	2262.6	2113.6	2463.6	0.0453	2.64
TP 4 - surface	94.50						
TP 4 - ?	89.05	31.6	1009.8	856.6	1127.5	0.0184	2.54

The frost heave susceptibility of the soils beneath the proposed summit house foundation are given in table 4. The values used to indicate the frost heave potential are calculated on the in-situ volume basis. The maximum ice expansion ratio (MIER) is defined as the ratio of the maximum ice volume to the soil void volume. The maximum ice volume is the volume that would be occupied by the moisture in the soil if it all became frozen. In practice, this never occurs as there is always some small fraction of soil moisture that does not freeze, the size of the fraction being dependent on the soil type. For example, the unfrozen fraction can be quite large for diatomaceous clays, but very small for coarse sands. However, the MIER is useful as an indicator of how close a particular sample is to reaching the frost heave condition. The largest MIER of .704 is shown for TP1 - 3, indicating that even if all the available soil moisture froze, there would still be about 30 % of the void volume left to accommodate additional moisture freezing. All the other sample locations show MIER's of less than .5 with the majority being less than .1. Hence these data show that the sub-foundation soils, if left in their undisturbed condition, are not susceptible to frost heave either from ice lens formation or from dilation as a result of the expansion of water upon freezing.

Field consolidation values also are reported in table 5. The field consolidation is a measure of how compressed (or consolidated) the soil has become in the

Table 5 Frost heave potential

Soil Sample Location	Maximum Ice Expansion Ratio	Field Consolidation (%)	Frost Produced Soil Dilation
TP1 - 1	0.098	43.64	no
TP1 - 2	0.049	12.84	no
TP1 - 3	0.704	59.26	no
TP1 - 4	0.216	29.97	no
TP1 - 5	0.064	-4.71	yes
TP2 - 1	0.101	25.79	no
TP2 - 2	0.028	-89.02	not determinable
TP3 - 1	0.038	-36.95	not determinable
TP4 - 2	0.432	32.56	no

field condition. It is expressed as the ratio between the decrease in saturated soil laboratory test void volume caused by field consolidation (that is, the test void volume minus the in-situ void volume) and the saturated soil laboratory test volume. Hence positive consolidation ratios show that the soil has been compressed in the field while negative values show soil dilation. Thus with the exception of TP1 - 5, TP2 - 2 and TP3 - 1, all the soil samples show varying degrees of consolidation, reaching a maximum of 59 % for TP1 - 3. Sample TP 1 - 5 shows a negative ratio of 5 % indicating that since TP1 - 5 was in virgin soil (not overburden), this location at some time has experienced frost heave or frost dilation. However, owing to its low MIER of .064, there is no expectation that such heave or dilation will occur again, provided the soil is left in an undisturbed condition.

In contrast, the large negative consolidations of TP2 - 2 and TP3 - 1 of -89 and -37 % respectively are not physically reasonable at depths of 5.9 ft or greater in undisturbed homogeneous ground, since upon excavation, these samples showed no obvious voids or indeed any other features indicating disturbance of their natural state. Hence in the absence of any other empirical evidence indicating substantial frost heave, it is not prudent to put much reliance on these data. Hence, the soil thermal properties for TP2 - 2 and TP3 - 1 to be used in the design simulation are taken from the data calculated on the water-saturated laboratory test volume basis. Further, this situation dictates that no statement on whether these soil samples have indeed experienced any frost related soil dilation can be made. However, noting their low MIER's even on the laboratory test volume basis (.076 for TP2 -2 and .062 for TP3 - 1) there is little probability that these locations will experience any frost heave in the future if left in their undisturbed state. The simulation soil properties specification is given in table 6.

Table 6 Computer simulation soil thermal properties

Soil Sample Location	Dry Density (kg/m³)	Porosity	Saturation Ratio	Dry Thermal Conductivity (W/m.K)
TP1 - 1	1953.3	0.239	0.539	0.391
TP1 - 2	1642.6	0.367	0.369	0.253
TP1 - 3	1996.3	0.228	0.894	0.418
TP1 - 4	1790.0	0.291	0.720	0.308
TP1 - 5	1497.3	0.421	0.434	0.210
TP2 - 1	1905.5	0.249	0.546	0.364
TP2 - 2	1484.9	0.426	0.476	0.207
TP3 - 1	1833.1	0.304	0.425	0.328
TP4 - 2	1644.1	0.352	0.838	0.254

These soil property data do not have any outstanding or unusual characteristics and show property values within the ranges measured for standard soils with similar particle size distributions. All the saturation ratios are less than the critical saturation ratio of .92 denoting the moisture content at which the soil is susceptible to dilation upon freezing. In other words, frost heave or ice expansion effects do not occur until the critical saturation ratio is reached. However, with saturation values of .894 and .838, samples TP1 - 3 and TP4 - 2 are within 0.026 and 0.082 of reaching the critical level, indicating that care must be taken in the foundation design to ensure that no additional water reaches these locations which could precipitate soil dilation. This is important even if no bearing loads are imposed, because any drilled columns or other structural elements passing through these regions would be susceptible to the imposition of adfreeze produced tensile loads. These loads arise when frost heave produced soil lifting forces are transferred to the column by shear through the soil/column adfreeze bond. Such structural elements could (and as a safety precaution, probably should) be designed to withstand some prudent level of adfreeze produced tensile loading

Closure

The importance of specifying soil properties in the thermal and structural design of building foundations is often overlooked, usually because such data are difficult to develop and the assumption that relative to other factors, they are of second order importance. The experimental earth-contact heat transport data provided by the FTF and illustrated briefly here, enables relevant computer simulation codes to be tested in terms of their ability to quantitatively replicate a variety of complex phenomena. In performing such testing, an adequate soil properties specification has been developed for use in these simulation codes. Two case studies showing examples of the soil property specification are invoked to provide quantitative illustrations of is use.

References

Comite Europeen. *Building Foundations - Protection Against Frost Heave*, Comite Europeen de Normalisation TC89 / WG 5, proposed CEN standard N198, preliminary draft, Brussels, 1992.

Farouki, OT. "Thermal Properties of Soils", *Series on Rock and Soil Mechanics*, Trans. Tech. Publications, vol. 11, 1986.

Goldberg, LF, Langenfeld, DT and Lively, RS. *Foundation Test Facility Experimental Results, Part I: 1993/94 Test Period Systems Data*, Underground Space Center, University of Minnesota, 1994.

Kersten, MS. "The Thermal Conductivity of Soils", *Proc. Highway Research Board*, vol. 28, pp. 391-409, 1948.

MN Dept of Admin. *MN Amendments to the 1988 Uniform Building Code*, MN Dept. of Administration, Building Codes and Standards Division, 1990.

NAHB *Design Guide for Frost-Protected Shallow Foundations*, prepared for the U.S. Dept. of Housing and Urban Development, NAHB Research Center, June 1994.

Status of ASCE Standard on
Design and Construction of Frost Protected Shallow Foundations

Larry S. Danyluk[1], M. ASCE, and Jay H. Crandell[2], M. ASCE

Abstract

A Frost-Protected Shallow Foundation (FPSF) is a practical alternative to deeper, more costly foundations in cold regions having seasonal ground freezing and the potential for frost heave. An FPSF incorporates strategically placed insulation to raise the frost depth around a building, thereby allowing foundations as shallow as 16 in., even in the most severe climates. This procedure has been used extensively in the Scandinavian countries over the last 40 years. ASCE is currently developing a Standard that would be used in the design of FPSFs. The Standard is based on proven Scandinavian practices and various studies performed in the U.S., including computer modeling and field verification tests. At the time of this paper, the Standard has gone through various iterations, but it is the intention of the Standards committee to have a document ready upon which they can vote by the end of this calendar year.

Introduction

Foundations in cold regions need to be adequately designed against frost action, otherwise damage may result from the heaving of freezing soil under and around them. In the past, foundations have extended below the depth of frost penetration in frost-susceptible soil. However, Farouki (1992) and Morris (1988) have reported that during the last 40 years there has been a tendency in the Scandinavian countries to use shallow foundations with associated insulation that utilizes some of the heat from the building to reduce the depth of frost penetration at the foundation. Figure 1 shows a Frost-Protected Shallow Foundation (FPSF) and a conventional foundation. Over 1 million FPSF homes have been successfully constructed in the Scandinavian countries.

---

[1]Research Civil Engineer, Civil Engineering Division, US Army Cold Regions Research and Engineering Laboratory, Hanover, NH.
[2]Director of Structures and Materials, National Association of Home Builders National Research Center, Upper Marlboro, MD.

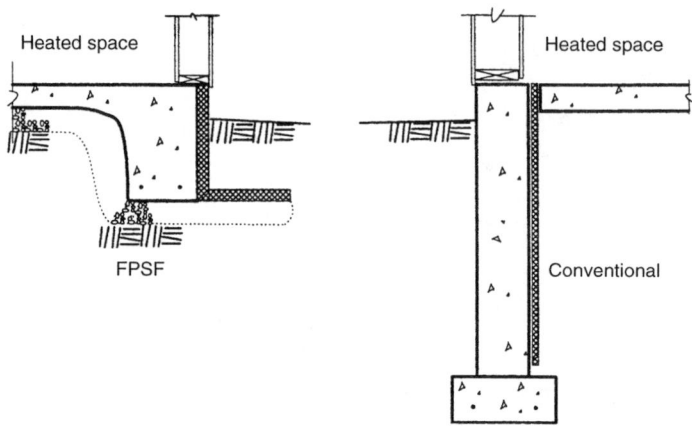

Figure 1. Schematic of FPSF and conventional foundation system.

The ASCE is currently developing a Standard to be used in the design of FPSFs. It does not apply to buildings on permafrost or to areas with mean annual temperatures less than 32°F. It applies to heated, unheated (T< 41°F), and semi-heated (41°F < T < 64°F) structures as based on the expected average monthly indoor temperature and type of building. The theory of using building and ground heat to prevent frost damage is briefly discussed in the next section. The paper further describes the process of generating the tables and figures used in the Standard, in particular the air freezing index map and field verification studies. The current status of the Standard will be discussed including problematic areas such as insulation, soil type, and crawl spaces. The European community is also trying to adopt a Comite European de Normalisation (CEN) standard, in collaboration with the International Standards Organization (ISO), on FPSFs and its status will be briefly addressed. The process of designing an FPSF using the ASCE Standard will be discussed, with special emphasis on the simplified design method.

Background

Historical Development. Morris (1988) reports that Scandinavian researchers believe that FPSFs were first used in a rudimentary way by early Nordic civilizations. Stone or sod walls laid directly on the ground surface were protected from frost heave by fires built inside the house and snow placed around the perimeter of the house to insulate the ground. In the U.S., slab-on-grade houses were constructed around the turn of the century in cold climates near Chicago. During the Depression, Frank Lloyd Wright designed and built a type of FPSF to meet affordability needs and used this technique in his "Usonian" style homes with shallow slab-on-grade foundations. The foundations were reported to be 6-in. deep, with hydronic heating system pipes laid on a bed of gravel beneath the slab.

During the 1950's Swedish and Norwegian researchers recognized the economic advantages of slab-on-grade construction and built various demonstration houses. In the 1970's the Scandinavian nations consolidated their research efforts in an attempt to address the implementation of FPSF technology. This ultimately lead to the publication in 1978 of the *Building Details* (NBI 1986). Advances in computer modeling technology have further defined the understanding of the parametric sensitivities of protecting shallow foundations with insulation. Variables such as extreme weather conditions, indoor temperature, temporary loss of building heat, and soil properties were investigated to extend practical experiences and insulation requirements in a numerical format conducive to the development of design guidelines.

Foundation Heat Flow. For a typical heated FPSF structure with slab-on-grade construction, about 10% of the heat loss from inside the building is through the floor (Torgersen 1976). The actual amount varies according to effectiveness of the insulation in the rest of the structure. The heat loss through the floor passes through the ground below and rises towards the outside air by means of an approximately semicircular path. The effect is to reduce the frost depth near the foundation compared with the frost depth in undisturbed ground. The heat flow from a floor tends to take the path with the least amount of thermal resistance, whether it be the connection between the floor and foundation wall (where a cold bridge could form), through the foundation wall, or through the soil under the floor and foundation.

In an FPSF structure, the insulation must be placed so that the majority of the heat loss through the floor is guided to the underside of the foundation. This can be done with floor, wall, and ground insulation, as shown in Figure 2. The type of insulation to use, how much, and where to place it depends on many variables including (but not limited to) climate, soil type, moisture conditions, building type, and use. Snow conditions and exposure are assumed in a worst case scenario (i.e., no snow or ground cover).

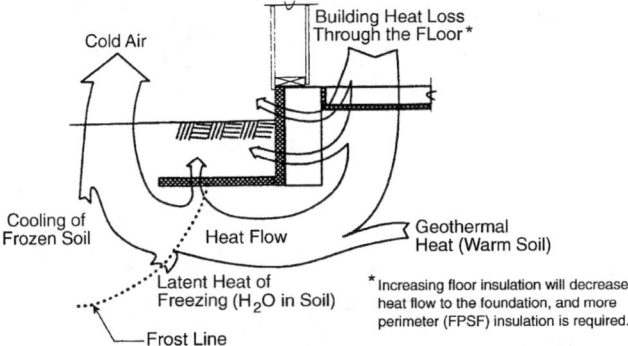

Figure 2. FPSF heat flow diagram for a heated building with optional floor insulation.

Unlike heated buildings, where heat flow from the building is utilized in foundation design, the protection of unheated, cold structures relies on the available soil heat that has been stored in the ground during summer. By using more insulation and placing it on a Non-Frost-Susceptible (NFS) drainage layer, damaging frost heave is prevented. Figure 3 shows an FPSF for an unheated versus a heated building. Foundation types such as wall (strip), column (point), and unheated areas of an otherwise heated building such as garages, exterior stairways, and breezeways are treated as unheated structures.

Advantages and Disadvantages. As with any construction technology, there are advantages and disadvantages. While the advantages will drive the desire to use FPSF technology, the disadvantages must be considered in an appropriate use of the technology. Some advantages include lower foundation costs (as much as 10% savings relative to conventional slab-on-grade foundations), reduced foundation construction cycle time, less excavation and site disturbance, and improved frost protection. Improved frost protection is provided by preventing frozen ground from adhering to or under-cutting the foundation. Since adhesion or adfreezing is often the cause of frost related damage in conventional masonry or concrete stem wall footings, the insulation acts as a "slip" surface and helps to prevent this, particularly in situations such as prolonged heating system failure. The insulation also serves as an effective thermal break between the soil below the building foundation and the outside environment.

Unheated

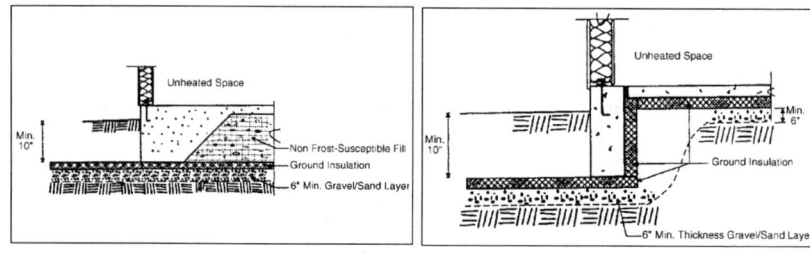

Heated

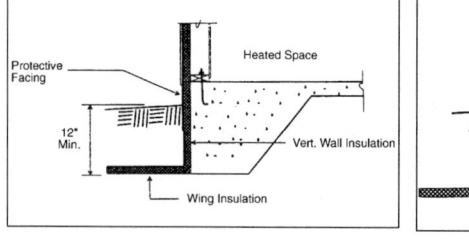
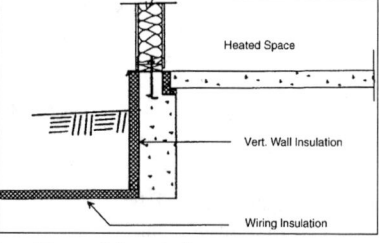

Figure 3. Schematic of an unheated and heated foundation system.

Some disadvantages include: 1) a building originally designed as heated cannot be later changed to a use that does not require interior conditioning, and 2) in the case of basement foundations, excessive amounts of insulation have led to higher lateral loads from soil freezing adjacent to the foundation because of reduced heat loss through the basement. Based on the experience in Europe and recent monitoring of FPSF performance in the U.S., it is believed that FPSF construction will be at least equivalent to current foundation frost protection methods (i.e., footings extended below a locally prescribed frost depth.) Superior performance (in terms of even further reduced risk of frost damage) can be obtained by designing the FPSF as though the structure is unheated, but not without significant construction cost increases.

Status of Standards

European Standard (CEN). Several years ago a working group under the auspices of the CEN standards committee (then a group of about nine European countries) formed for the purpose of consolidating knowledge on FPSFs. The collective experience and research on FPSFs was assembled as a pre-standard document that was submitted to the CEN approval process (Comite European de Normalisation 1996). A parallel submission has also been made to the International Standards Organization (ISO). These processes are expected to be complete in the fall of 1997. The foundation insulation requirements in the final draft of the CEN standard are exactly those in the draft pre-standard for the ASCE standards committee.

ASCE Standard. The first formal meeting of the ASCE standards committee was held in July 1996. The committee originally had 19 members, and has since expanded to 24. Membership on the committee is divide into three groups—producers, consumers, and general interest. The committee is considered in balance when any one group has at least 20% but not more than 40% of the total committee membership. At the time of this paper, the committee is in balance.

One of the first decisions made by the committee was to use the CEN/ ISO Standard and the *Design Guide for Frost-Protected Shallow Foundations* prepared by NAHB Research Center (1994) as the primary references in developing the draft ASCE standard. Some additional information to assist in proper design and code compliance in the U.S. has been added. Also, details compatible with U.S. construction practices are substituted.

The committee has meet three times and, at the time of this writing, is working on the third draft of the standard. Three areas have generated a lot of discussion— insulation, soils, and crawl spaces. The first two topics have brought about the formation of subcommittees to further investigate the subjects. At the time of this writing, crawl spaces will be treated as a heated or semi-heated space and will be designed as such. The crawl space will be unvented and must be adequately protected from moisture by the use of an approved vapor barrier on the ground surface inside of it and by providing adequate site drainage. As more information becomes available these specifications may change.

The insulation subcommittee was tasked to develop recommendations on insulation adjustments (effective resistivities) and factors of safety for insulation loadings. Since insulation materials are subject to compression from soil and building loads, the insulation shall have adequate compressive strength and deformation characteristics to resist the loads during the life of the structure. Compressive loads supported by the insulation shall not exceed the compressive strength reported in ASTM C 578-92, divided by a safety factor yet to be determined (ASTM 1992). The nominal (dry) insulating value of the material shall be modified to reflect its performance in the moist conditions encountered in a foundation application. The effect of moisture on the effective resistivity of insulating materials is a controversial issue; there is no universally accepted testing method to measure thermal resistivities that duplicate conditions experienced by buried insulation. This, in turn, raises the question of which insulating materials are acceptable for buried applications, particularly in moist or horizontal applications. The subcommittee hopes to have these issues settled by the completion of the next draft.

The soils subcommittee addressed the issue of the selection of the "characteristic" soil for use in the modeling of insulated foundations. The soil properties were selected on the basis of sensitivity studies using analytical heat transfer models. These models then used the characteristic soil to determine the insulation requirements needed to prevent frost damage. The subcommittee also decided that any backfill material placed next to or under an FPSF will be considered frost-susceptible if the amount passing the #200 sieve exceeds 6%.

Development of ASCE Standard

Tables and Figures. The majority of the figures and graphs used in the Standard are from CEN (1996). The ASCE committee members felt that, since the CEN Standard represents the consensus of the European community and that this foundation system has performed as expected in thousands of applications, there was no need to "reinvent the wheel." The committee felt it was more important to generate new AFI maps for the U.S. and to verify that the values in the tables and graphs are compatible with the climate we have here. Verification would be done by field demonstration sites and extensive monitoring.

Air-Freezing Index. The Air-Freezing Index (AFI) is a major component of a FPSF design in that it is a measure of the combined magnitude and duration of temperature above and below freezing during any given winter season. When the AFI is combined with the soil properties, it is possible to determine the ground-freezing potential for any given site or soil condition. Steurer and Crandell (1995) found that there are many different methods of generating freezing indices, and the differences in these methods have a substantial influence on the estimated AFI and the application of return-period statistics.

Steurer and Crandell found that the three most common methods of computing the AFI are those used in the U.S., Norway, and Finland. The U.S. method is defined as the cumulative departure of the mean daily temperature above and below 32°F between the highest and lowest points on a cumulative degree-day curve for one freezing season. The index is calculated on a seasonal basis that spans a 12-month period. The Norwegian method parallels a commonly used index known as heating degree-days. The AFI is the daily cumulative sum of the departure of temperatures below 32°F; any temperatures above 32°F are disregarded. The Finnish AFI is defined as the departure of the monthly mean temperature below 32°F, multiplied by the number of days in the month for each month in the winter season.

It became obvious that the Finnish and Norwegian methods overestimated the AFI in that they negate above-freezing thawing events. Furthermore, the length of time over which the temperature is averaged can produce widely different results, particularly in moderate climates such as what we have in the U.S. Steurer and Crandell (1995) concluded the U.S. method to be most representative of the freezing effect for all U.S. climates because it captures all freezing and thawing events during the winter seasons, regardless of their duration or time of occurrence. (The European community is currently switching over to the U.S. method.)

A typical FPSF is designed using a return period of 1 in a 100-year AFI. Steurer and Crandell (1995) found that the Weibull distribution generated the best fit AFI using readily available 30-year temperature records. They compared their predicted results with long term temperature records. Figure 4 is an AFI contour map in °F-days obtained by using temperature data from over 3000 stations and fitting it to a Weibull probability distribution to produce an estimate of the 100-year return period.

Field demonstrations. Demonstration structures were designed and constructed in Alaska, Vermont, Iowa, North Dakota, and Oregon using design data extensively from the CEN draft standard, Crandell et al. (1993, 1994), Danyluk and Khosrownia (1995) and Danyluk (1997). All the buildings were designed for a 100-year return period. Collectively, these sites offered a variety of geographic and climatic situations, including different soil types, topographies, and winter climates. Although there have been many other FPSFs successfully built, these foundations were instrumented with temperature and moisture sensors to continuously monitor and assess their performance and determine which parameters have the most impact on the overall performance of the foundation. All the foundations were slab-on-grade; however, some used monolithic construction techniques and the others used a short stem wall. Both extruded polystyrene (XPS) and expanded polystyrene (EPS) insulations were used in various configurations and thicknesses.

All the sites (Crandell et al. 1994, Danyluk 1997) had winter seasons with freezing indices within 11% of the average (2-year mean return period) and, as a result, all of the foundations performed as expected. The freezing front never threatened to get beneath the bottom of the footing. In some of the buildings, the apparent severity of the winter was increased because the buildings were not completed until the late fall or early winter, well after the start of the heating season. Thus, the structure provided no contribution toward building an equilibrium of heat reserves in the ground before the

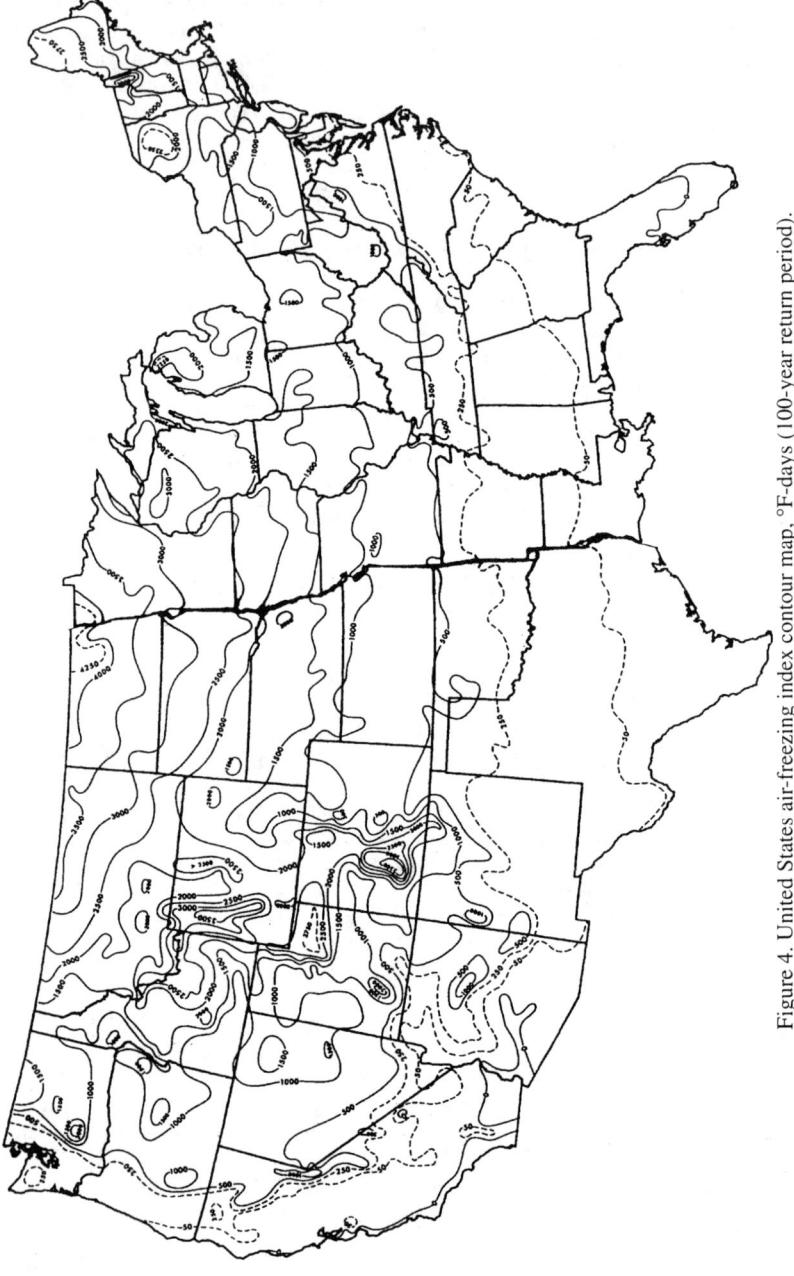

Figure 4. United States air-freezing index contour map, °F-days (100-year return period).

onset of winter; also, in some cases, the indoor temperatures were kept significantly lower than normal since the buildings were unoccupied. As a realistic result, the monthly temperatures were low at times, which in turn created a more stringent test condition. Since the FPSF design of the heated parts of a building relies on heat flow through the floor, this demonstrated the capability of the FPSF even when indoor temperatures are lower than expected.

Crandell et al. (1994) wanted to further test FPSFs under extreme environmental conditions. They accomplished this by designing the foundation using the normal 100-year return period with the exception of a test zone created in each design. The insulation in the test zone was sized for an *average* winter event instead of the normal 100-year return. In effect, a near average winter event would simulate a 100-year return period test of the frost protection provided by the design guidelines. Although the winter season was slightly more severe than average, the frost line was successfully stopped from penetrating into the subgrade underneath the perimeter of the building in the test zone. The frost penetration was significantly less in the remaining areas of the foundation, where normal levels of insulation were utilized for frost penetration. This finding indicates that the insulation requirements specified in the design guidelines provide a high degree of protection against frost heave.

Another aspect of the demonstration sites was to determine the constructibility and cost effectiveness of the FPSF. Although FPSFs require no special construction techniques, some of the contractors needed to become familiar with constructing monolithic slabs. However, once this was accomplished, the foundation construction proceeded more rapidly and efficiently. One contractor estimated that it was two to four times faster to construct an FPSF than a conventional foundation. All test sites reported a savings of at least 10% of the foundation costs, with some reporting savings of over 50%. This translates to a savings of 1–5% of the total cost of the structure.

Design Procedures

The ASCE Standard is divided into four main design sections: simplified design for heated buildings, detailed design for heated buildings, design for unheated buildings, and special design conditions. Because of the limited scope of this paper, each section is only briefly described below. To give the reader an example of the types of tables and figures used in a FPSF design, the simplified design section is slightly expanded. For more detailed information, the reader is advised to see CEN (1996), NAHB National Research Center (1994), or the ASCE draft Standard.

Simplified design for heated buildings. In consolidating the design steps for the simplified method, thermal resistance values for the vertical insulation were established so that the performance level of various conditions, including slab surface temperature, were conservatively accommodated. To use the simplified approach, the AFI for the site location must be known. Insulation resistivity values and dimensions, and the depth of the footings, are then determined from Table 1 and Figure 5. When foundation depths are greater than 12 in. are required by Table 1, the increase in depth may be satisfied by substituting gravel, crushed rock, sand, or approved NFS material.

More economical construction costs may be obtained when the detailed design procedure is used. The inherent conservative approach of the simplified design adds somewhat to the cost of the foundation. The detailed design procedure must be used when buildings include unheated areas, such as attached garages.

Detailed design procedures for heated buildings. In practice, there are many different combinations of vertical and horizontal insulation details, thermal resistivity values, and footing depths that can be used in an FPSF. The detailed approach is a flexible approach that allows the designer to utilize experience and select the preferred method of construction for a given site. For example, the designer may opt to provide vertical wall insulation only, wing insulation only at the corners, or provide wing insulation around the entire building. The designer also has the flexibility to step the footing to increase foundation depths, add wing insulation to reduce foundation depths, or select the width of wing insulation in meeting the minimum requirements in the design process. The design approach for crawl-space foundations is similar to the approach used for slab-on-grade buildings. Many different combinations of vertical and horizontal insulation can be used in an FPSF. These insulation values are a minimum for frost protection; additional insulation could be used to increase energy efficiency.

Table 1. Minimum Insulation Requirements for
Frost-Protected Footings in Heated Buildings[1]

Air-Freezing Index (°F days)[2]	Vertical Insulation R-Value[3,4]	Horizontal Insulation R-Value[3,5]		Horizontal Insulation Dimensions per Figure No. 5 (inches)			Minimum Footing Depth (inches)
		along walls	at corners	A	B	C	D
1,500 or less	4.5	NR	NR	NR	NR	NR	12
2,000	5.6	NR	NR	NR	NR	NR	14
2,500	6.7	1.7	4.9	12	24	40	16
3,000	7.8	6.5	8.6	12	24	40	16
3,500	9.0	8.0	11.2	24	30	60	16
4,000	10.1	10.5	13.1	24	36	60	16

[1]Insulation requirements are for protection against frost damage in heated buildings. Greater values may be required to meet energy conservation standards. Interpolation between values is permissible.

[2]See Figure 4 for Air-Freezing Index values.

[3]Insulation materials shall provide the stated minimum R-values under long-term exposure to moist, below-ground conditions in freezing climates. The following R-values shall be used to determine insulation thicknesses required for this application: Type II expanded polystyrene—2.4R per inch; Types IV, V, VI, VII extruded polystyrene—4.5R per inch; Type IX expanded polystyrene—3.2R per inch. NR indicates that insulation is not required.

[4]Vertical insulation shall be expanded polystyrene insulation or extruded polystyrene insulation.

[5]Horizontal insulation shall be extruded polystyrene insulation.

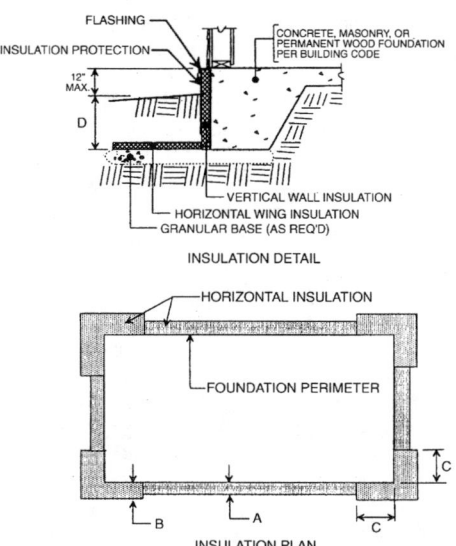

Figure 5. FPSF design parameters for heated
buildings using the simplified design procedure.

While the heated building design procedure is based on a monthly average indoor temperature of greater than 64°F, the design approach is modified for the semi-heated condition. A structure is considered semi-heated if the average monthly temperature is between 41 and 64°F and the design procedure is to simply follow the design for a heated condition, except for increasing the foundation depth by 8 in.

Design method for unheated buildings. The design of unheated foundations relies on geothermal heat reserves. For this reason, a mat of insulation is placed on the ground below the entire building footprint. The designer has the flexibility to increase foundation depth to reduce ground insulation requirements.

Special design conditions. In buildings that have both heated and unheated areas, special considerations are required. Depending on the size of the unheated area, the design may require a deeper foundation or additional insulation, or both. When large unheated areas are encountered, the heated and unheated sections are regarded as separate buildings and the insulation is designed accordingly. In buildings that are semi-heated, the building is designed as if it is a heated building but the foundation depth is increased by 8 in.

Insulation must be protected from ultraviolet radiation, physical damage, or other sources of deterioration, such as petroleum based products. Also, buildings in areas favorable to termite infestation and constructed of materials susceptible to termite damage shall be protected against termite infestation by approved methods, such as mechanical barriers and soil treatment.

Conclusion

In the United States, FPSF technology has been used in engineered structures and is common in some parts of the country, but the major model building codes do not specifically recognize the FPSF's equivalence to footings placed below a prescribed frost depth. Recent amendments to the CABO (1992) *One and Two Family Dwelling Code* recognize the simplified design. The BOCA (1990*) National Building Code* and the ICBO (1991) *Uniform Building Code* recognize performance-based criteria or alternate approaches for frost protection, but do not specifically mention FPSFs. Consequently, widespread use of the technology is possible, particularly with engineering designs for commercial structures or prescriptive designs for homes. It is hoped that with the establishment of an ASCE standard on FPSF, combined with building code acceptance, the potential of FPSFs will be fully realized.

References

American Society for Testing and Materials. *Standard Specification for Rigid, Cellular Polystyrene Thermal Insulation*. ASTM C 578–92, Philadelphia, PA (1992).

Building Officials and Code Administrators (BOCA) International, Inc. *National Building Code*. Country Hills, IL (1990).

Comite European de Normalisation (CEN). *Building Foundations—Thermal Design to Avoid Frost Heave*. Final draft for proposed European Standard N502E, CEN/TC89/WG5 (January 1996).

Council of American Building Officials (CABO). *One and Two Family Dwelling Code*. Falls Church, VA (1992).

Crandell, J.H., Lund, E.M., Bruen, M.G. and Nowak, M.S.. *Frost Protected Shallow Foundations, Phase I*. Prepared for the U.S. Department of Housing and Urban Development by NAHB Research Center, Inc., Upper Marlboro, MD (April 1993).

Crandell, J.H., Lund, E.M., Bruen, M.G. and Nowak, M.S.. *Frost Protected Shallow Foundations, Phase II*. Prepared for the U.S. Department of Housing and Urban Development by NAHB Research Center, Inc., Upper Marlboro, MD (June 1994).

Danyluk, L.S. *Shallow Insulated Foundation at Galena, Alaska, a Case Study*. Special Report 97-7. USA Cold Regions Engineering and Research Laboratory, Hanover, NH (1997).

Danyluk, L.S. and Khosrownia, G. *Shallow Insulated Foundations for Pre-Engineered Metal Buildings*. Corps of Engineers Structural Engineering Conference, San Antonio, Texas (August 1995).

Farouki, O. *European Foundation Designs for Seasonally Frozen Ground,* USA Cold Regions Research and Engineering Laboratory, Hanover, NH, Monograph 92-1 (1992).

International Conference of Building Officials (ICBO). *Uniform Building Code*. Whittier, CA (1991).

Morris, R.A. *Frost-Protected Shallow Foundations: Current State-of the Art and Potential Application in the United States.* NAHB Research Center, Inc., Upper Marlboro, MD (1988).

Norwegian Building Research Institute (NBI). Slab-on-grade with foundation wall: Heated buildings. *Building Details*, Norwegian Building Research Institute, Worksheet A521.811 (1986).

NAHB Research Center. *Design guide for Frost-Protected Shallow Foundations.* Prepared for the US Department of Housing and Urban Development by NAHB Research Center, Inc., Upper Marlboro, MD (1994) .

Steurer, P.M. and Crandell, J.H. *Comparison of Methods Used to Create Estimate of Air-Freezing Index.* ASCE Journal of Cold Regions Engineering, p. 64–74 (1995).

Torgersen, S.E. Frost protection of floor on groundwall. *Frost I Jord*, **17**(10): 287–314 (1976).

Aspects of Geotechnical Engineering in Permafrost Regions

James M. Oswell[1] & Alan J. Hanna, M.ASCE [2]

Abstract

Geotechnical engineering in permafrost regions have many aspects in common with temperate geotechnical engineering, but also many aspects that are unique due to the climate, geological history and often remoteness of sites. This paper provides a discussion of some of those unique aspects, providing a general overview of foundation systems used to support structures where the ground is perennially frozen. Several case histories are discussed that illustrate some problems of permafrost engineering.

Introduction

In North America, the arctic and subarctic regions cover some 12.5 million square kilometres. Within this land mass lie the State of Alaska, the Yukon and Northwest Territories of Canada, and parts of four Canadian provinces. The temperatures range from -50°C to over +35°C. Whereas in the lower forty eight states the ground may experience seasonal freezing of the soil, and the conventional foundation design is to place the bearing level below the depth of seasonal freezing, in the permafrost regions, engineers are faced with seasonal thawing, where the ground is perennially frozen, and only the upper several metres may thaw in the summer.

---

[1] Senior Permafrost Engineer, Manager of Geotechnical Division. AGRA Earth & Environmental Limited. 221 18 Street, S.E., Calgary, Alberta, CANADA, T2E 6J5

[2] Senior Permafrost Engineer, Manager of Northern Operations. AGRA Earth & Environmental Limited. 221 18 Street, S.E., Calgary, Alberta, CANADA, T2E 6J5

Whereas frost is the worry to temperate engineers in the south, it is thawthat is the worry to permafrost engineers. The design of foundations in permafrost regions is based on the modified adage, "If it ain't thawed, don't thaw it".

The design and construction of foundations in this region is completely unlike anything in the more temperate zones of the continent. In this paper, an over view of foundation systems used in permafrost regions is provided. In addition, other aspects of geotechnical engineering in cold regions are provided. It is not the intent that this paper represent a state-of-practice, but rather to touch on some of the broader issues involved in the subject.

The important parameters for unfrozen soils to the design of foundations are the undrained strength for cohesive soils, or the effective stress parameters for all soils, depending on the particular application. Natural water content, in situ density are incorporated into the unfrozen strength parameters. Unfrozen soils are a three phase system (mineral grains, water and air) but often modelled as a two phase system (mineral grains and water).

In frozen soils the composition of the soil is often more complex. In many cases, there are at least four phases or components (mineral grains, ice, water, and air). In addition, the water phase can be complicated by the presences of salts that depress the freezing point. The strength properties of the soils are temperature dependent. As a general rule, coarse grained sands and gravels tend to contain low excess ice (that is, ice in excess of frozen water to completely fill the void spaces between the mineral grains). Fine grained silts and clays tend to contain considerable excess ice, especially in the upper 3 to 5 m. Ice can also exist in bedrock, most commonly in sedimentary rocks and also in cracks and fissures in any rock.

The impact of ground ice on foundations can be significant. Even if the ground is preserved in a frozen state, soil with excess ice exhibits long-term deformations (creep) under sufficiently high sustained loads. This limitation in the load bearing capacity of ice-rich permafrost is particularly significant where the mean annual ground temperature is warmer than - 5 C or if the pore ice is saline. Excess ice in soils may also produce significant settlements if the soils are permitted to thaw. Such soils are termed "thaw sensitive". Soils that do not exhibit significant settlement on thawing are termed "thaw stable".

Some northern areas of Canada have been subject to significant isostatic rebound (upward rebound of the earth's crust after the retreat of the ice sheets). Thus areas that were once undersea now form the coastal regions with numerous small towns and villages. The effect of the rebound is that many coastal communities are underlain by deposits that have salinity approaching that of sea water (Hanna, 1989). The result of this salinity is a reduced strength in the bond between saline ice and mineral particles, and that the load deformation behaviour is markedly different from fresh water soils (Nixon and Lem, 1984).

In the past twenty five years, the use of numerical models to predict the performance of structures in and on permafrost has increased significantly. Ho, Harr and Leonards (1970) and Goodrich (1974) published numerical models for the analysis of geothermal problems.

It is not intended to address geothermal modelling applications in this paper. The reader is referred to texts by Andersland and Anderson (1978) and Lunardini (1981, 1991) for more information on this subject.

Foundations on Permafrost

Foundations for buildings in permafrost may be divided into two broad categories: on-grade and elevated. The choice of foundation type is usually controlled by the type of structure. Buildings with very heavy floor loads or requiring at-grade access such as maintenance garages or fire halls are typical of on-grade foundation systems. Other buildings, such as offices, residences, and schools are typically designed as elevated structures.

On-grade Structures

The design approach for these buildings is based on heat interception. The following list ranks systems in decreasing commonality, increasing reliability and in most cases, increasing cost:

- naturally ventilated and insulated pad
- forced ventilated and insulated pad
- insulated pad
- cooling pipes in pad (refrigerated or thermosyphons)
- heat pump circulation in pad

Figures 1 and 2 show a variety of on-grade structures that are commonly used in cold regions.

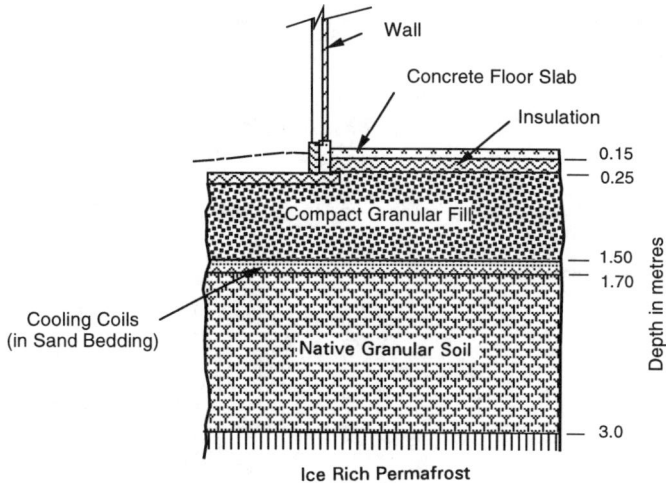

Figure 1.   Typical On-grade Foundation Configuration with Artificial Cooling

The geothermal requirements for the design of on-grade structures include the amount of insulation, spacing of ducts, heat pipes and thickness of the granular pad. The required inputs are the structure temperature, the floor/fill configuration (for access), and the mean ground temperature at depth. Nixon (1978a) has prepared an approximate solution for this problem.

Natural ventilation has been a traditional method of intercepting building heat and introducing cold winter air to the foundation pad. The ventilation pipes were usually orientated to the prevailing winds. This system has a number of drawbacks, including snow drifting that blocks the cold winter air entering the pipes, and the need to seal the pipes in the summer to prevent warm air from entering. To overcome these concerns, the pipes were fitted with vertical stacks to induce a chimney effect (Odom, 1983). The required diameter of the pipes to promote natural ventilation is typically in the order of 450 to 600 mm.

Forced ventilation systems are more reliable, but have a higher capital and maintenance cost. The principle is the same as natural ventilation, expect the vertical stacks include a fan to force cold winter air through the piping. Thermostats control the fans to ensure warm air is not circulated. With forced ventilation systems, the diameter of the ducts can be reduced to about 150 mm, thus reducing the total fill thickness beneath the slab.

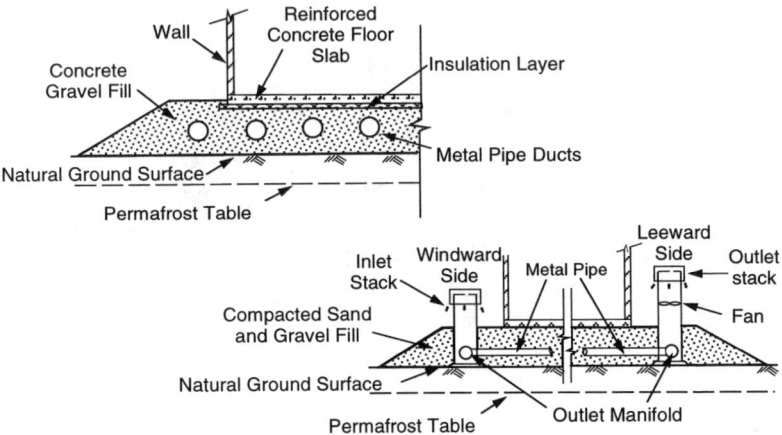

Figure 2.   Typical On-grade Foundation Configurations with Ventilation Piping

On-grade structures that rely solely on insulation are also feasible. They may be used where the permafrost is cold (less than about -8°C), as it is necessary for the permafrost to act as sink for any heat passing through the insulation. Nixon (1983) prepared a design chart for simple problems.

Cooling systems that rely on active refrigeration have been used in a number of special applications. The recreation centre in Inuvik, Northwest Territories was constructed on a site with a surficial layer of peat, underlain by ice-rich permafrost. In order to prevent the building heat from warming the permafrost, the refrigeration system used for the curling and skating rinks was also used to cool the subgrade during the summer.

Thermosyphons have been successfully used to maintain frozen conditions below on-grade structures in Canada and Alaska (Hayley, 1982, 1988; Zarling et al., 1990). These devices rely on a "low" temperature evaporant to draw heat from the subgrade and transfer it to the atmosphere; they operate whenever the ground temperature is warmer than the air temperature.

Elevated Structures

The design concept in considering an elevated structure is to provide an air space between the building and the ground surface that will be sufficient to ensure building heat does not warm the permafrost and produce long-term thawing. For most buildings, an air gap of 0.6 m is sufficient, although for buildings larger than 3000 m^2 in area, an air gap of 1.0 m is recommended. Figure 3 shows a two storey building supported on timber piles with an air gap.

Figure 3.   Multi-storey Building Supported by Adfreeze Timber Piles

There are several types of foundations used to support elevated structures. The foundation choice is often based on both economic and technical considerations. Considering the remoteness of some communities in the far north, the equipment may not be available to construct a particular type of foundation. Ground temperature, ice content and salinity of the sub soils must be considered in the design process. The main types of foundation systems are footings and piles. Piles may be further subdivided into adfreeze, grouted, or thermal piles.

Footings

Footings placed on the permafrost level can provide sufficient support for many single and two storey buildings. Considerations in the design of such foundations include the following factors:

- soil type, ground temperature, salinity and ice content
- depth of pre-construction active layer
- predicted depth of post-construction active layer
- insulation requirements
- frost jacking protection for footing column
- timing of construction

- bearing capacity
- settlement
- maintenance requirements

For sites warmer than about -2°C, it may not be economical to prevent long term degradation of the permafrost. Therefore, it would be necessary to confirm that the sub soils are thaw stable; that is, that they do not contain excess ice that will produce settlement on melting. Sites colder than -2°C are sufficient to support footings, although insulation may be needed around the footing to provide long-term protection of the permafrost, and may be also used to raise the permafrost level to permit the placement of shallower footings.

The required amount of insulation typically ranges from 25 mm to 75 mm depending on the ground temperatures. Figure 4 illustrates a typical design chart for a particular mean ground temperature (-5°C) that relates depth of the footing to the amount of insulation, based on geothermal modelling.

Bearing capacity of footings on permafrost is a function of allowable long-term creep settlement, unless the soils are thaw stable. Nixon (1978b) addresses the bearing capacity of footings. The limiting creep rate should consider the warmest temperatures that may be experienced by the footings, typically within a depth below the footing of two footing widths. When designing on the basis of creep rate, the corresponding "ultimate" bearing capacity is reduced by 1.5 to determine the allowable bearing capacity. The relatively low factor of safety is used because the creep rate method is itself inherently conservative.

Piled Foundations

Figure 3 illustrates a large building supported on timber piles. The use of piles in northern Canada has a relatively short history (fifty years) and their design and construction methodologies are still evolving. Piles in Alaska have been used since the turn of the century (Crory, 1963). Significant advances in piling technology resulted from the construction of the Town of Inuvik in the late 1950's where over 20,000 timber piles were installed, the design and construction of the Alyeska pipeline in 1975 where over 60,000 piles were required to support the above ground section, and during the development of the Alaskan north slope.

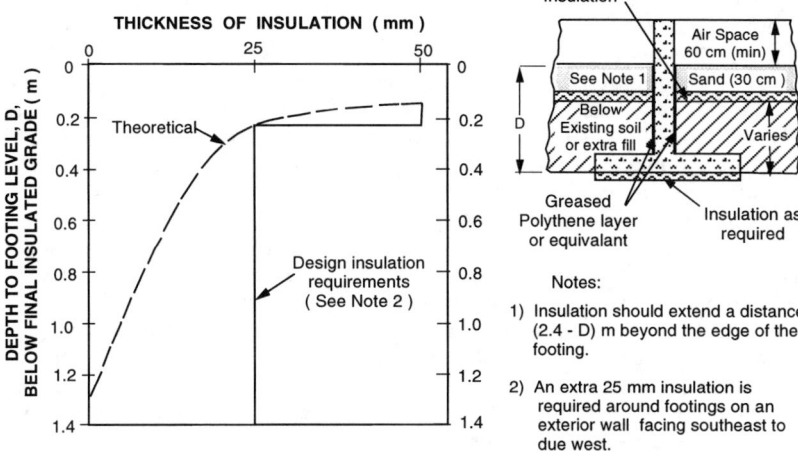

Figure 4. Design Example of Insulation for Shallow Footing

Piles in permafrost may be divided into several categories: adfreeze piles, grouted piles or driven piles. The selection and design of piles in permafrost depends on many variables such as:

- magnitude of loads and allowable settling rate
- type of soil, ice content, salinity, and ground temperature
- the available adfreeze and shear strength of the soils
- pile type and available materials and equipment
- economics

Equipment for pile installation is a very important consideration when designing pile foundations in the north. In many isolated communities the entire construction equipment inventory may consist of one air-track drill rig, one front end loader, one rubber tired backhoe and perhaps one grader or bulldozer. Furthermore, the available size of the drilling equipment is often limited, to either 144 mm or 164 mm diameter holes. Hence it is impractical to specify pile diameters that are not consistent with the drilling capabilities or specify pile lengths that are physically impossible to install.

Adfreeze Piles

Adfreeze piles rely on the bond between the pile material and a frozen slurry of water and sand to provide bearing capacity. They are typically constructed by drilling an oversized pile hole, partially backfilling the hole with unfrozen slurry, forcing the steel pipe, or timber pile through the slurry to the base of the hole and then filling the remaining annulus with slurry. The slurry then freezes forming a bond between the pile and the native soil.

The adfreeze bond between frozen soil and a smooth steel pipe may be considerably less than the shear strength of a frozen soil. Newcombe (1973) and Vialov (1959) suggest that the adfreeze bond strength is only 15 to 30 percent of the frozen shear strength. More recent laboratory testing has shown that the failure surface in adfreeze piles is the pile/backfill interface (Sego and Smith, 1989). The type of backfill material and water quality also influence adfreeze and bearing strength. Markin (1973) reported that when clay rather than sand was used in the slurry, a 33 percent decrease in load capacity was experienced. Tsytovich (1973) confirmed the merits of using medium grained sands. Saline free sand and water should be used in the production of the slurry. When salts are present, a reduction in the adfreeze bond must be considered due the freezing point depression of the water. For example, Nixon (1988) related pile loads and settlement rates to the salinity, as shown in Figure 5.

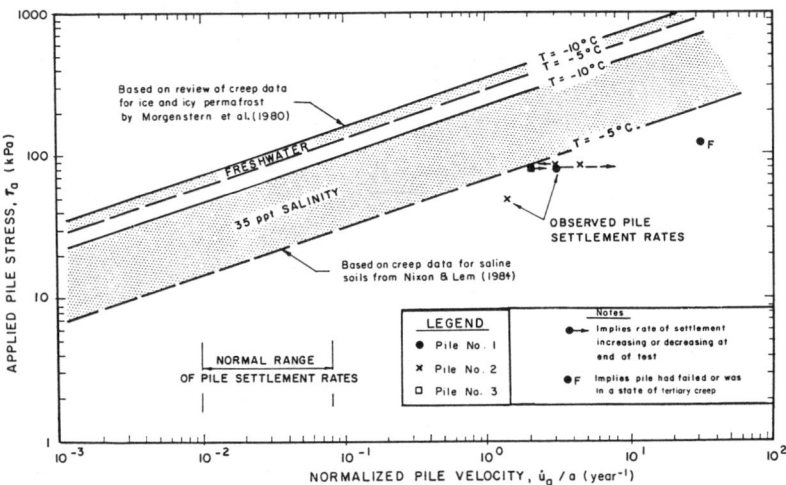

Figure 5.   Applied Pile Stress versus Pile Settlement for Adfreeze Piles (from Nixon, 1988)

The adfreeze pile design is governed by the most critical of the following: the strength between the frozen soil and pile surface (usually critical), the shear strength of the frozen slurry/backfill, or the shear strength of the native soils. The long-term strength of the frozen soil has been studied in laboratory shear strength and model testing, and field pile load tests. The basis of adfreeze pile design has been described by Nixon and McRoberts (1976), with further contributions by Morgenstern et al (1980), and Weaver and Morgenstern (1981). The settlement of piles in permafrost have been converted to a normalized creep velocity that is considered relative to the allowable shaft adhesion, as shown in Figure 6 from Weaver and Morgenstern (1981).

Grout backfilled piles may also be used as adfreeze piles. In this case, the pile is grouted into a hole drilled in permafrost soils, and the pile hole diameter, rather than the pile diameter is used to calculate the capacity.

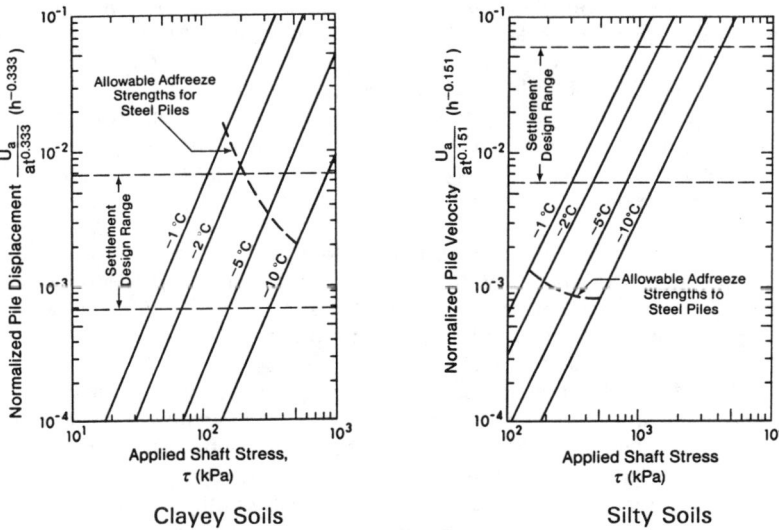

Clayey Soils   Silty Soils

Figure 6.   Design Charts for Adfreeze Piles for Clays and Silts (from Weaver and Morgenstern, 1981)

Grouted End Bearing Piles

As much of the eastern arctic of Canada is underlain by shallow bedrock, grouted piles are often a viable option. They tend to have a higher load capacity than adfreeze piles if the bedrock is sound and the full grout strength can be developed. In sound rock, the pile capacity may be limited by the yield stress of the steel or the confined strength of the rock.

The grout used in permafrost regions is a special blend usually being a high alumina cement or containing magnesium phosphate. The intent of these cements is to hydrate very rapidly and exhibit high early strengths even at sub freezing temperatures. Compressive strengths of 46 MPa in laboratory specimens after 24 hours curing at temperatures of -9°C to -7°C were reported by Nawy et al. (1987). Specimens cast by the authors from grout prepared by a contractor on site in Baffin Island during the installation of grouted piles exhibited a 28 day compressive strength of about 40 MPa after curing at temperatures of about -10°C.

Driven Piles

There has been some success with the installation of driven piles in permafrost. Most of the applications have been in either Alaska, or Russia. Crory (1963), Gosstroi (1969), Nottingham and Christopherson (1983), and Manikian, (1983) discuss the use of these piles. Limited use has occurred in Canada, particularly in the western Arctic in recent years. Timber piles were the first type of pile used, but steel H sections or hollow steel sections are now used in North America, while pre-cast concrete piles are popular in Russia.

The piles are often driven into undersized boreholes. The design capacity may be based on Weaver and Morgenstern (1981) or Nixon (1988). In ice -rich soils caution must be taken to apply sufficiently low adfreeze stresses in order to limit the rate of creep settlement.

Care should be practised when using pile load test data from piles driven in undersized holes. The long-term adfreeze strength may be smaller at a future date than obtained from load tests conducted shortly after installation. Lateral load relaxation and warming of permafrost due to surface disturbance may decrease the adfreeze strength with time. These effects should be taken into account when interpreting pile load test data and in foundation design.

Thermal Piles

Thermal piles are piles in which natural convection or forced cooling systems have been installed to remove heat from the ground. The thermal devices are normally used with sand slurry adfreeze piles in warm permafrost. They are used to decrease the freeze back time, prevent long-term degradation of the permafrost and decrease the existing ground temperature around adjacent piles. Thus pile capacities can be greater than those based on the natural ground temperatures.

The most common type of thermal piles are the natural convection systems referred to as "thermosyphons", "thermo tubes", "convection cells" or "heat pipes". Thermosyphons and heat pipes consist of a two-phase heat transfer loop that removes heat from the soil by vaporization and releases it to the atmosphere by condensation in a sealed tube. Gas/liquid materials that have been used in the tubes are propane, ammonia and carbon dioxide. The devices operate in the winter when the ground temperature is warmer (causing the liquid to vaporize and consume heat) than the air temperature (which causes the vapour to condensate and release the heat to the atmosphere). During the summer months the liquid remains dormant in the base of the tube.

The devices are sealed and generally maintenance free, however, inspection by the manufacturer every few years is considered prudent. They are subject to damage and vandalism, particularly the above ground portions.

Impacts of Building Construction on Permafrost

In most situations the interaction between adequately designed and constructed buildings and the surrounding or supporting ground is benign. Settlement rates are usually within allowable tolerances and the buildings perform generally as desired. In this section, three case histories are reviewed to show the impact of development.

Camp at Tuktoyuktuk

The Gulf Canada Resources camp at Tuktoyuktuk, Northwest Territories was constructed in 1983. Figure 3 shows a view of the main camp building, which is supported on timber piles. The camp was the operations base for the company as it pursued oil and gas exploration in the Canadian Beaufort Sea. The camp was effectively shut down in the early 1990's, although the company has taken care to ensure the facility remains in good repair.

The authors were requested to investigate a problem of building heave. In some places, (external stairs) the camp buildings appeared to have heaved by large amounts, in some cases more than 300 mm. In the investigation that ensued, hand augers were used to investigate the upper zone of the active layer. The surficial soils at the site consisted of a granular pad, underlain by peat, underlain by low plastic silty clays. The active layer under the buildings was typically 75 percent of the active layer outside the buildings.

If the buildings were in fact heaving, the movement was very uniform, for none of the dozens of windows in the building had cracked, nor had the drywall cracked indicating little differential movement (many of the doors were difficult to close or open, but the occurrences were completely random, and believed to be due to environmental rather than structural causes). The camp maintenance operations were also reviewed. It was reported that during the camp's operating period, the areas around the buildings were cleared of snow in the winter, and that this had continued after the camp was closed, except for the last several years. For the past winters, no snow clearing was done as there were few visitors to the camp to justify the expense.

This comment was the key to mystery. Rather than the buildings heaving, it was considered that the camp area, particularly the open areas were settling. The geothermal regime of the area had been altered by the construction. While the camp was operating, the snow clearing permitted the cold winter temperatures (-40°C) direct contact with the ground. Hence, without the benefit of the insulating snow layer, the granular pad and underlying soils experienced colder than normal temperatures. During the summer, the summer heat would be readily absorbed by the gravel pad, warming the subsoils. However, the cooling effect in winter was greater than the warming effect in summer.

When snow clearing stopped in the early 1990's, the energy balance shifted. The summer energy absorbed by the gravel pad and subsoils was the same, but in the winter, the snow blanket insulated the ground from the severe cold, and the ground did not cool much. The net affect was that the thaw front would progress deeper, likely encountering ice rich soils that were now thawing and experiencing thaw settlement.

Hay River Coast Guard Building

The shop and stores building for the Coast Guard was built in 1967 over approximately 7 m of ice-rich silty clay permafrost underlain by unfrozen till deposits. The structure was supported on pile foundations with a structural floor. By 1974 some settlement of the soils beneath the building was reported. Although no structural damage was occurring, the soils around the perimeter of the building was starting to slough under the perimeter grade beams. Up to 1.5 m of sand fill was placed under the floor in 1974. An extension to the building was also added in 1974.

A investigation in 1980 indicated that the permafrost had degraded on the south side of the building. On the north side, approximately 2.5 m of permafrost remained at depth. Thermistor strings installed by the authors' company in 1988 confirmed that all subsoils were unfrozen

(Hanna et al, 1990).

The amount of thaw settlement to be experienced by a soil is directly related to the volumetric water content (Hanna, 1988). Data collected over a number of years is presented in Figure 7. These soils are generally thaw stable where the volumetric water content is less than about 38 percent. Based on the original water content (in 1968) and the initial thickness of permafrost, the predicted settlements ranged from 1.8 m to 2.1 m. This ignores the consolidation settlement following thawing.

Actual observations around and under the building showed evidence of settlements exceeding 1.5 m.

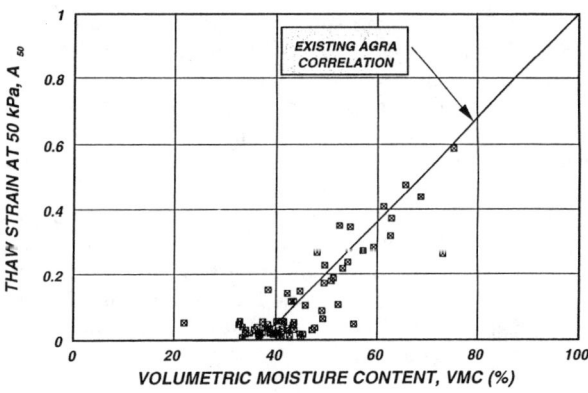

Figure 7. Thaw Settlement Response for Soils in Permafrost Regions

Dawson City 4-Plex

Trimble and Jacobsen (1996) reported an interesting building problem in Dawson City. The building, in the first year following construction was experiencing considerable distress, including cracking drywall, and sticking doors and windows. This was despite the fact that the building was supported on piles grouted into bedrock. The overlying soils consisted of ice-rich permafrost. Consistent with good permafrost practice the building was elevated about 0.6 m above the ground so that the building heat would not thaw the ice-rich permafrost.

A survey program was initiated to monitor the building and foundation movements. The survey data confirmed that the building was indeed moving, but the mystery was that the foundation system appeared to be stable. Considerable effort was spent on trying to reconcile the two conflicting performances

The problem was finally solved when it was recognized that the building had a 2 by 4 stud hoarding around the building, that covered the free air space. The hoarding had likely been built in the summer months, was constructed tight to the ground. During the winter months, the active layer in re-freezing would experience some heaving, which in turn raised the hoarding, and lifted the structure off the foundation. As Trimble and Jacobsen report "the problem was quickly solved" with the aid of a chain saw.

Underground Services

The construction, maintenance and performance of underground services can also be a challenge in permafrost soils. In Iqaluit (Frobisher Bay), on Baffin Island, and several other communities in the eastern Arctic, the sewer and water services are directly buried in the permafrost. The lines are insulated with urethane foam insulation. Some sewer lines have suffered collapse, within several years following installation.

In 1989, the authors were requested to investigate the geotechnical aspects related to problems with the buried HDPE sewer lines in Iqaluit, Northwest Territories. The community is located on the southeast corner of Baffin Island. The problems included various forms of pipe collapse, crushing of pipe insulation, and other distortional features of the pipe coatings and access (manhole) vaults (Hanna and Cucheran, 1991, 1994).

Because the community is located in the zone of continuous permafrost, with a mean ground temperature of about -5°C, the sewer lines are insulated with up to 75 mm of urethane insulation. The insulation is protected by an outer extruded HDPE jacket. The HDPE pipe varies from Series 45 through Series 160.

An investigation was initiated to identify the causes of pipe failures in the Town's sewer main system. Excavations were made at locations with collapsed pipe, identified by physical blockages or by sewer video cameras. The observations suggested two different types of pipe failure. They were: failure due to pressure external to the outer jacket, and failures due to pressures inside the outer jacket. Figure 8 show the typical pipe shapes for each type of failure mode.

Both types of failure are believed to be due to excessive hydrostatic pressures developed during freeze back of the active layer. The fact that in some cases, the pressures appeared to have originated inside the outer jacket reflects a problem of poor seals at pipe joints. Considerable pressure can occur around the pipes if a closed system develops during freeze back. Closed freezing refers to the case where the freezing zone around the pipe is confined in both the transverse and longitudinal directions. As freezing continues inward towards the pipe the normal volume increase in freezing soils induces the pressure on the pipe.

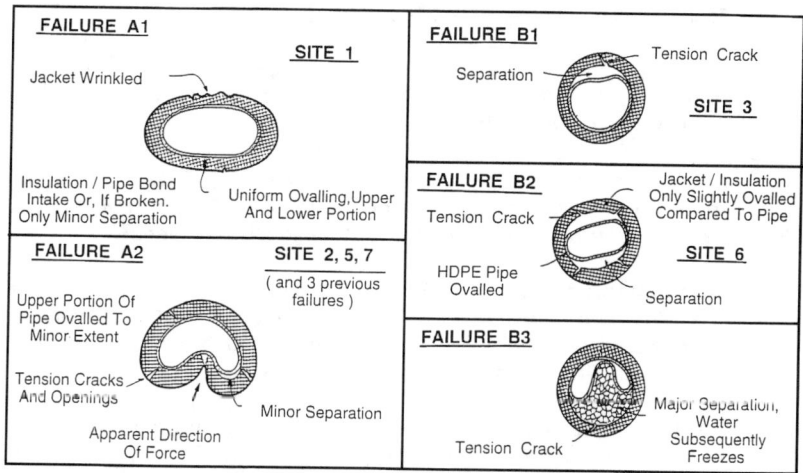

External Pressures   Internal Pressures

Figure 8.   Deformed Shapes of Sewer Pipes due to Internal or External Pressures

Conclusions

Geotechnical engineering in northern Canada, Alaska and other permafrost regions of the world involve a wide variety of problems not encountered in more temperate climates. The harshness of the environment often imposes strict limitations as to what may be technically or economically achievable. The environment, remoteness and economic barriers all act to promote innovation by the engineers. It may be true that there is no such thing as a routine job.

References

Andersland, O.B., and Anderson, D.M. (editors), (1978). *Geotechnical engineering for cold regions*. McGraw Hill Book Company, Toronto.

Crory, F.E. (1963). "Pile foundation in permafrost." *Proceedings, 1st International Conference on Permafrost*, Lafayette, Indiana.

GOSSTROI. (1969). "Handbook for the design of bases and foundations of buildings and other structures on permafrost. *Technical Translation, Research Institute of Bases and Underground Structures, Gasstroi, U.S.S.R.* Translated by Poppe, V., NRC Technical Translation TT - 1865 (1976).

Goodrich, L.E. (1974) A one-dimensional numerical model for geothermal problems. *National Research Council of Canada, Division of Building Research*. Technical Paper 421.

Hanna, A.J., Forsyth, R.J., and Garvin, D. (1990). "Thaw settlement around a building on warm ice-rich permafrost." *Proceedings, 5th Canadian Permafrost Conference*, Quebec City, Quebec.

Hanna, A.J., and Cucheran, J., (1991). "Problems with buried sewers in The Town of Iqaluit, N.W.T." *Proceedings, 10th Canadian Hydrotechnical Conference/Engineering Mechanics Symposium*, Vancouver, British Columbia.

Hanna, A.J., and Cucheran, J., (1994). "Monitoring and remediation of problems with the buried sewer system in the Town of Iqaluit, N.W.T.". *Proceedings, 7th International Cold Regions Engineering Specialty Conference*, Edmonton, Alberta.

Hayley, D.W. (1982). "Application of heat pumps to the design of foundations in permafrost." *Proceedings, 4th Canadian Permafrost Conference*, Calgary, Alberta.

Hayley, D.W. (1988). "Maintenance of a railway grade over permafrost in Canada." *Proceedings, 5th International Conference on Permafrost*, Trondheim, Norway.

Ho, D.M., Harr, M.E., and Leonards, G.A. (1970). "Transient temperature distribution in insulated pavements: Predictions and observation." *Canadian Geotechnical Journal*, 7, 275 - 284.

Lundardini, V.J. (1981). "*Heat transfer in cold climates*". Van Nostrand Rheinhold. New York.

Lundardini, V.J. (1981). "*Heat transfer with freezing and thawing*". Van

Nostrand Rheinhold. New York.

Manikian, V. (1983). "Pile Driving and load tests in permafrost for the Kuparuk pipeline system." *Proceedings, 4th International Conference on Permafrost*, Fairbanks, Alaska.

Markin, K.F. (1973). "Bearing capacity of piles in permafrost." *Proceedings, 2nd International Conference on Permafrost*, Yakutsk, Russia.

Morgenstern, N.R., Roggensack, W.D., and Weaver, J.S. (1980). "The behavior of friction piles in ice and ice-rich soils." *Canadian Geotechnical Journal*, 17, 405 - 415.

Navy, E.G., Hanaor, A., Balaguru, P.N., and Kudlapur, S. (1987). "Early strength of concrete patching materials at low temperatures." *Concrete and Concrete Construction, Transportation Research Record 1110, Transportation Research Board (NRC)*, 24 - 33

Newcombe, T. (1973). "B.P. Alaska pile test program." *B.P. North Slope Project*, Anchorage, Alaska.

Nixon, J.F. and McRoberts, E.C. (1976). "A design approach for pile foundations in permafrost." *Canadian Geotechnical Journal*, 13, 40 -

Nixon, J.F., (1978a). "Geothermal aspects of ventilated pad design." Proceedings, *3rd International Conference on Permafrost*, Edmonton, Alberta.

Nixon, J.F. (1978b). "Foundation design approaches in permafrost areas." *Canadian Geotechnical Journal*, 13, 96 - 112.

Nixon, J.F. (1983). "Geothermal design of insulated foundations for thaw prevention." *Proceedings, 4th International Permafrost Conference*, Fairbanks, Alaska.

Nixon, J.F., and Lem, G. (1984). "Creep and strength testing of frozen saline fine-grained soils". *Canadian Geotechnical Journal*, 21, 518 - 529.

Nixon, J.F. (1988). "Pile load tests in saline permafrost at Clyde River, Northwest Territories." *Canadian Geotechnical Journal*, 25, 24 - 32.

Nottingham, D., and Christopherson, A.B., (1983). "Driven piles in permafrost: State of the Art." *Proceedings, 4th International Conference on Permafrost*, Fairbanks, Alaska.

Odom, M.E. (1983). "Practical application of underslab ventilation system: Prudhoe Bay case study." *Proceedings, 4th International Conference on Permafrost*, Fairbanks, Alaska.

Sego, D., and Smith, L. (1989). "Effect of backfill properties and surface treatment on the capacity of adfreeze pipe piles." *Canadian Geotechnical Journal*, 26, 718 - 725.

Trimble, R. and Jacobsen, N. (1996). "Experience with foundation designs in the Yukon Territory." *Proceedings, 49th Canadian Geotechnical Conference*, St. Johns, Newfoundland.

Tsytovich, N.A. (1973). *The mechanics of frozen ground*. Translated from Russian., Swinzow, G.K. and Tschebotarioff, G.P., editors. Scripta/McGraw Hill, New York.

Vialov, S.S. (1959). "Rheological properties and bearing capacity of frozen soils." *Cold Regions Research and Engineering Laboratory, U.S. Army Corps of Engineers Translation 74.*

Weaver, J.S., and Morgenstern, N.R. (1981). "Pile design in permafrost." *Canadian Geotechnical Journal*, 18, 357 - 370.

Zarling, J.P., Hansen, P., and Kozisek, L. (1990). "Design and performance experience of foundations stabilized with thermo-syphons." *Proceedings, 5th Canadian Permafrost Conference*, Quebec City, Quebec.

Spatial and Temporal Variability of Estimated Maximum Soil Freezing Depths in the Northeastern U.S.

Arthur T. DeGaetano and Daniel S. Wilks[1]

Abstract

A physically-based model for estimating frost penetration is used to develop a time series of soil frost penetration extremes. Since the model is based on meteorological variables which are observed at U.S. Cooperative Network stations, frost depth statistics can be estimated for approximately 500 sites within the northeastern United States.

Using model simulations, the performance of a suite of theoretical probability distributions for representing the resulting distributions of extreme soil freezing depths is investigated, and it is concluded that the Gumbel distribution provides the best regional representation of soil freezing extremes. This distribution is used to develop maps showing the spatial distribution of maximum frost depths corresponding to several return periods. In addition, the season-to-season and station-to-station variability in annual maximum frost depth are examined across the northeastern United States.

Introduction

Information regarding the presence and maximum depth of soil freezing is necessary for a variety of climate-sensitive applications ranging from building design to agricultural operations. Given this diversity in applications, requests for soil freezing data are frequently received by the Northeast Regional Climate Center (NRCC). Unfortunately, the only published source of this information that is apparently available is based on unofficial, undocumented and antiquated (1899-1938) measurements (USDA, 1941). This situation has led us to develop a one -dimensional heat flow model similar to that described by Benoit and Mostaghimi (1985). The resulting physically-based model estimates the rate and depth of frost penetration under bare soil and sod using only temperature, precipitation and snow depth data. Since records of these data extend from the late 19th century to the present and given that approximately 8000 stations are in operation nationwide (about 900 are in the northeastern U.S.), model-derived frost depth statistics can be developed for a relatively dense network of sites.

[1] Sr. Research Assoc. and Assoc. Professor, respectively, Northeast Regional Climate Center, Cornell University, 1115 Bradfield Hall, Ithaca, NY 14853

Description of Model

Ideally, the calculation of regional soil freezing statistics should incorporate the features of physically-based models to the extent possible (e.g. Flerchinger and Saxton 1989; Guymon et al. 1993; Cary et al. 1978). Given the limitations of meteorological measurements from a relatively dense network of stations, the use of such models in this application is not feasible. Alternatively, more empirical models (e.g. Aldrich and Paynter 1953) tend to require less detailed meteorological data, but incorporate assumptions that make their use unrealistic for applications in which snow cover varies with time. The model that is briefly described in this section combines the desirable features of these two classes of soil freezing models. A more complete description of the model is given in DeGaetano et al. (1996).

The theoretical basis for the model is that the process of frost penetration is driven primarily by thermal diffusion. Figure 1 illustrates this principle. At the lower boundary, Z_D, which is set at a depth of 2 m, a daily "deep" temperature T_D is given as a function of the average air temperature over a period from the previous April through March of the current year, the 25th percentile January through March snow depth for the current year and the combined thermal diffusivity of the snow and soil. The model assumes that the flux of heat through the lower boundary is negligible.

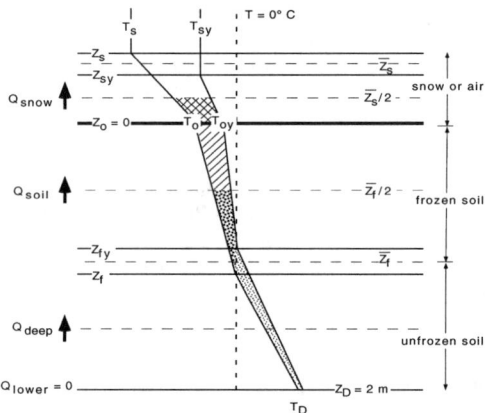

FIGURE 1. *Schematic diagram showing the model's frozen soil state. Depths below or above (in the case of snow and/or air) the surface are indicated by Z and temperatures are indicated by T. Subscripts indicate snow (s), frozen soil (f), the soil surface (0), and the lower boundary (D). The subscript "y" refers to a value observed or estimated for the previous day. Heat fluxes through the centers of each layer are indicated by bold arrows. The stippled area represents the change in energy storage* ΔQ_L *and the hatched areas represent* ΔQ_U.

The upper boundary condition is given by the observed average daily air temperature. Here the assumption is made that the average daily air temperature is representative of the temperature of the snow surface. The snow depth (Z_s) gives the thickness of the first layer in the snow/soil system (Fig. 1). In the absence of snow cover, the air temperature is assumed to equal the temperature at the upper surface of a 1.0×10^{-3} m laminar layer, the thermal properties of which are characteristic of still air. Progressing downward, soil layers of variable depth are defined by frozen and unfrozen zones, the boundaries of which are at $0°C$ (Fig. 1). A maximum of three soil layers (one frozen and two unfrozen) is allowed by the model.

Temperature gradients through each layer are assumed to be constant, and thus the heat fluxes at the middle of each layer, Q_x, are defined by the differences between the temperatures of the layer boundaries. Imbalances between the resulting vertical heat fluxes (i.e. heat flux convergence or divergence) are rectified through internal temperature changes and, when these changes cross $0°C$, freezing or thawing of an appropriate depth of soil. In this process, the fluxes are balanced by accounting for the heat capacities of soil solids and soil water, and for the latent heat of fusion. This is sketched in Figure 1 and given mathematically for the case when a frozen layer exists at the surface by the governing equation

$$Q_{snow} = Q_{froz} = Q_{deep}, \tag{1}$$

where

$$Q_{snow} = -K_{snow}(T_s - T_0)/Z_s + \Delta Q_U, \tag{2}$$

$$Q_{froz} = K_{froz}(T_0 / Z_f) + \Delta Q_L, \tag{3}$$

and

$$Q_{deep} = K_{thaw}T_D /(Z_f - Z_D) + (Z_{fy} - Z_f)(\varepsilon - 0.1)L_f. \tag{4}$$

The variables used in Equations 1-4 are defined in Figure 1 with the exception of the latent heat of freezing (L_f), soil porosity (ε), thermal conductivities of snow (K_{snow}), frozen soil (K_{froz}), and unfrozen soil (K_{deep}) and the change in heat storage terms (ΔQ). Equations 1–4 are solved numerically for the prognostic variables T_0 and Z_f. Nearly-saturated soil moisture conditions are assumed at all times. In Figure 1, ΔQ_U is represented by the hatched and cross-hatched areas between the two consecutive daily average temperature profiles. Similarly, ΔQ_L is shown by the speckled and dotted regions.

Only one of three possible soil-freezing states is illustrated by Figure 1. In this state a layer of frozen soil extends from the surface to some depth, Z_f. The other possible states are that the soil may remain unfrozen from the surface to the lower boundary, Z_D, or a layer of frozen soil may exist between two layers of unfrozen soil. In addition, five transition modes are possible, corresponding to the transitions between the three basic states with the exception of the transition from unfrozen to a buried frozen layer, which is not physically realizable.

The model is initiated in the unfrozen state and continues in this manner until T_0 falls below $0°C$. At this point, the transition to frozen soil mode is activated. Provided the temperature remains below $0°C$ on subsequent days, the model operates in the frozen soil mode. In this state, both soil freezing and thawing occur at the bottom of the frozen layer. When T_0 exceeds $0°C$ the model transitions to either the unfrozen or surface thaw state. In the surface thaw state, the layer of frozen soil is allowed to thaw both from its top and bottom. The temperature throughout the buried frozen layer which results is assumed to be a constant $0°C$. For subsequent occurrences of $T_0 < 0°C$ freezing occurs at the both the top and bottom of the buried frozen layer. Although physically unrealistic, in that freezing should be allowed to occur at the top of the surface-thaw layer, this formulation allows a solution without liberal assumptions regarding the temperature profile within a subsurface thawed layer. Given that the purpose of our model is to estimate the depth of maximum frost penetration, the omission of a second frozen layer at the surface is of little consequence.

Model Verification

Validation of the model was possible using frost depth data collected with six Army Corps of Engineers frost depth tubes (Ricard et al., 1976). These gauges were installed at the Ithaca, New York weather observation site, allowing coincident soil-freezing and daily meteorological data to be collected. Three gauges were placed under sod. The remaining gauges were located within a bare soil plot. Frost depth measurements were made on a weekly basis. Verification for bare soil case during the winter of 1995-1996 is presented in Figure 2.

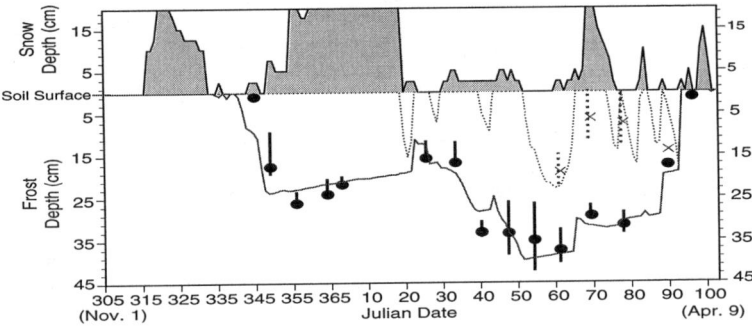

FIGURE 2. *Observed versus estimated frost depths under a bare soil surface at Ithaca, NY for the winter of 1995-96. Modeled frost depths and surface thaw depths are given by the solid and dashed lines, respectively, and correspond to y-axis values ≤0. The bars show the range of three frost depth observations, with bold dots identifying the median. Dashed vertical lines and Xs similarly indicate observed thawing. Only two frost depth measurements were available on Julian days 38, 59, 66, 75 and 87. Daily snow depths are indicated by the gray regions at the top of the plot*

During the winter of 1995-96, observed maximum frost depths under bare soil ranged from 42 to 26 cm, with a median frost depth of 34 cm measured by the third gauge (Fig. 2). In all cases, the maximum frost depth occurred between 21 and 29 February. During this winter, the maximum model-derived value of 40 cm occurred on 20 February indicating exceptional correspondence between the observations and model estimate in both the timing and magnitude of maximum soil freezing. Over the course of the winter the model tracks the timing and progression of soil freezing quite closely (Fig. 2). Over the 15 soil freezing observations shown in Figure 2, the model exhibits a 1.8 cm bias toward overestimation of the observed frost depth, with a mean absolute error of 3.4 cm. Similar validation results were obtained for the sod-covered surface.

Maximum Frost Depth Climatology

Data

Annual model-derived maximum frost depths were calculated for a set of 306 northeastern U.S. cooperative network stations (Fig. 3). To be included, stations were required to have at least 30 years of non-missing daily climatological data. At all sites serially complete daily temperature data (DeGaetano et al., 1995) was available for one of four periods (1951-1993; 1951-1990, 1961-1993 or 1961-1990). Serially complete snow depth and precipitation data were not available. If missing, these parameters were estimated based on values at neighboring stations. Years were not considered if any missing data values occurred for more than 7 consecutive days during October through April.

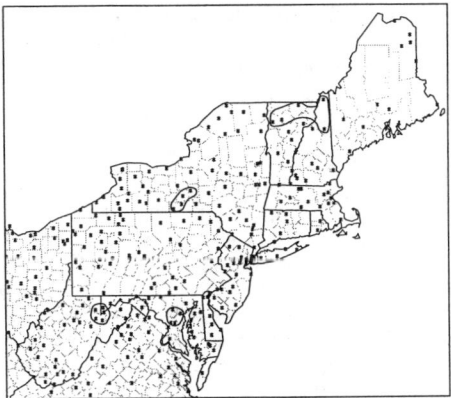

FIGURE 3. *Locations of stations used to develop the extreme soil freezing climatology. At least 30 annual maximum frost depth values are available at each site. Outlined areas encompass stations used for spatial and temporal variability analyses.*

Computation of return periods

Smoothing and extrapolation of the modelled annual maximum frost depth data for all stations was accomplished by fitting the Gumbel distribution (Wilks, 1995). This distribution was selected from a pool of 11 candidate distributions using a bootstrap procedure described by DeGaetano et al. (1997) The probability density function for the Gumbel distribution is

$$f(x) = \frac{1}{\beta}\exp\left\{-\exp\left[-\frac{(x-\xi)}{\beta}\right]-\frac{(x-\xi)}{\beta}\right\}, \tag{5}$$

where x is the random variable (in this case, annual maximum frost depths). The distribution has two parameters: ξ is a location parameter, and β is a scale parameter. Separate distributions are fit to the data for each station by maximum likelihood. One convenient feature of the Gumbel distribution is that it is analytically integrable, so that its cumulative distribution function can be written in closed form. That is, Gumbel probabilities can be obtained using

$$F(x) = \Pr\{X \le x\} = \int_0^x f(x)dx = \exp\left\{-\exp\left[-\frac{(x-\xi)}{\beta}\right]\right\} . \tag{6}$$

Average return periods, R, relate to cumulative probabilities, F, of the distributions of annual maximum data according to

$$R = \frac{1}{\omega[1 - F(x)]} , \tag{7}$$

where ω is the average sampling frequency, in this case 1 yr^{-1}. Subsequently, frost depths, x, corresponding to specified return intervals are obtained by solving Equation 6 for x and substituting the expression $F(x) = 1 - 1/R$, obtained by rearrangement of Equation 7. These operations yield the expression for frost depths as a function of return period and the parameters of the fitted Gumbel distribution,

$$x = \xi - \beta \ln[-\ln(1 - \frac{1}{R})]. \tag{8}$$

Return period mapping

Maps depicting the spatial distributions of maximum frost depths for specific return intervals were prepared by first gridding the individual station values, and then producing contour maps from the gridded fields by automated means. The details of this gridding and contouring procedure are given in DeGaetano et al. (1997). As an example, Figure 4 shows a map of the 100-year return periods for maximum annual frost depth under bare soil using observed snow cover conditions. Similar maps can be generated for sod-covered and snow-free bare soil conditions. In addition, maps depicting 2-, 5-, 10-, 25- and 50-year return periods for these three surface conditions are available from the authors.

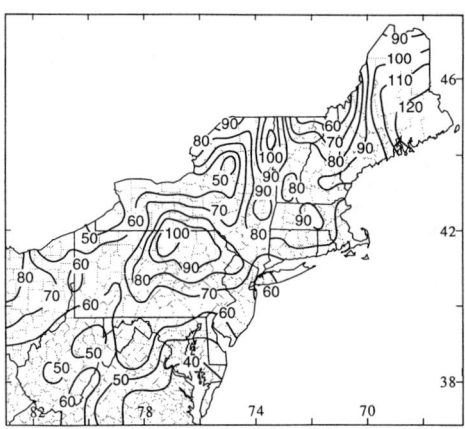

FIGURE 4. *100-year return period frost depths (cm) under bare soil.*

Temporal variability

Annual maximum frost depth time series for several of the 306 sites shown in Figure 3 were analyzed with regard to their temporal characteristics. Distributions of annual maximum frost depths under bare soil over the period 1951-1995 for four stations are shown by boxplots in Figure 5. These sites, chosen based on their geographic and climatological diversity, are included within the outlined areas in Figure 3. Within each subregion, the year-to-year variability in maximum soil freezing depth is considerable. However, similar temporal variation exists among the subregions. For instance, the 0.14 m interquartile range at the southernmost Maryland site, is similar to that found at the West Virginia and New York sites. Year-to-year variations in frost depth are higher at the New England site as evidenced by the 0.22 m interquartile range.

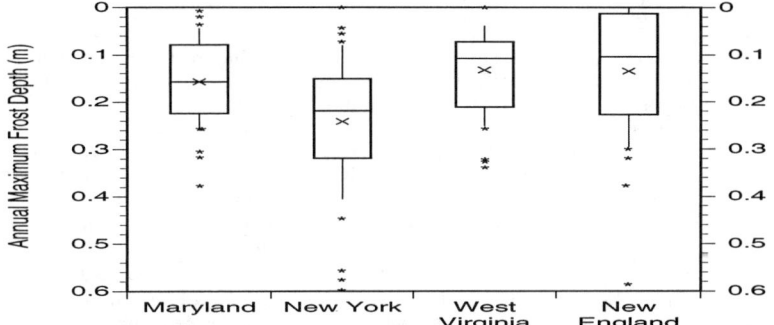

FIGURE 5. *Boxplots of annual maximum soil freezing depths (m) during the period 1951-1995 for individual stations in four climatically diverse regions.*

In general, the annual frost depth series exhibit a positive skewness, that is the majority of data points lie below (i.e. are shallower than) the mean value (means are indicated by an × in Figure 5). This is expected as the frost depths are constrained by zero and the mean tends to be inflated by the occurrence of extremely deep soil freezing levels in isolated years. However at the Maryland station, the distribution of frost depths is more symmetric about the mean. Interestingly, frost-free winters were observed in each subregion , with the exception of Maryland. In fact, the New England station experienced freeze-free conditions during more than 10% of the winters. Presumably this is a reflection of ample snow cover at the more northern (and mountainous) sites.

The time series in each region were also analyzed for time-dependent trends. Figure 6 shows the annual series for two representative sites, Glenn Dale, Maryland and Newport, Vermont. Based on least squares linear regression, each station exhibits a tendency toward shallower frost depth with time. However neither of these trends is statistically significant.

Spatial variability

The spatial variability of annual maximum soil freezing depth is perhaps of more practical interest than temporal variations. Given a soil freezing map such as

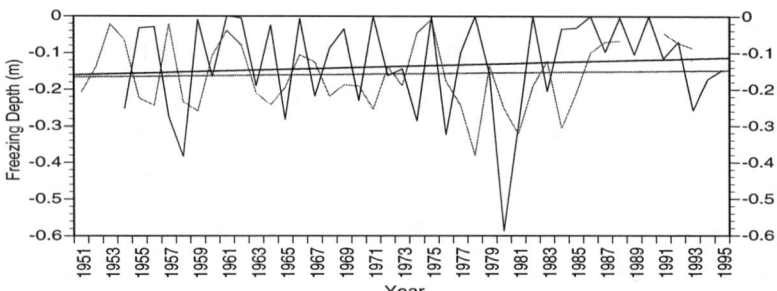

FIGURE 6. *Annual maximum soil freezing depth time series for Glenn Dale, Maryland (gray) and Newport, Vermont (black) with least squares regression trend lines superimposed.*

that presented in Figure 4, it is often necessary to interpolate frost depth contours or actual station data to site-specific locations. Thus, it is useful to quantify the spatial variability in annual maximum frost depth that arises due to small-scale (microclimatic) differences in climatological conditions as opposed to the regional differences that are depicted in Figure 4. To evaluate these spatial differences, four groups of stations were selected such that the geographic distance (and climatic variability) between stations within each group was relatively small. However, individual groups encompassed the major climatic and geographic regimes of the northeastern U.S. These station groupings are outlined in Figure 3.

In general, the deepest maximum frost depths are found at the northernmost sites. In central New York and the mountains of northern New England the deepest observed frost depths exceed or approach 1 m (Fig. 7). At the southern stations, maximum depths are on the order of 0.5 m. However at individual stations (shown by separate bars in Figure 7), similar maximum observed frost depths occur within each subregion. In fact, among the four subregions the shallowest observed maximum frost depths (and the shallowest 50-year return period depths) vary by as little as 18 cm. Within each subregion, except for the area in eastern Maryland, between-station differences in maximum observed frost depth exceed this between-subregion value. Clearly site-specific characteristics have a pronounced effect on annual maximum soil freezing depth.

Figure 7 summarizes several factors that may contribute to this spatial variability in soil freezing. With the exception of the eastern Maryland subregion, it appears that between-station differences in snow depth account for the majority of the spatial variation in frost depth within each subregion. Figures 7a-c, show a strong tendency for frost depth to increase as average January-February snow depth and annual maximum snow depth decreases. In the eastern Maryland subregion, snow cover is relatively uncommon at all sites leading not only to a diminished relationship between soil freezing depth and snow depth, but also relatively little between station variation in soil freezing depth. A clear relationship between elevation or temperature and frost depth is not apparent in any of the four subregions.

Based on the results shown in Figure 7, it is quite apparent that the existence and persistence of snow cover can have a profound effect on the spatial (and temporal) variability of soil freezing. To isolate the effects of temperature variations on the depth of soil freezing, the maximum soil freezing depths given in Figure 7

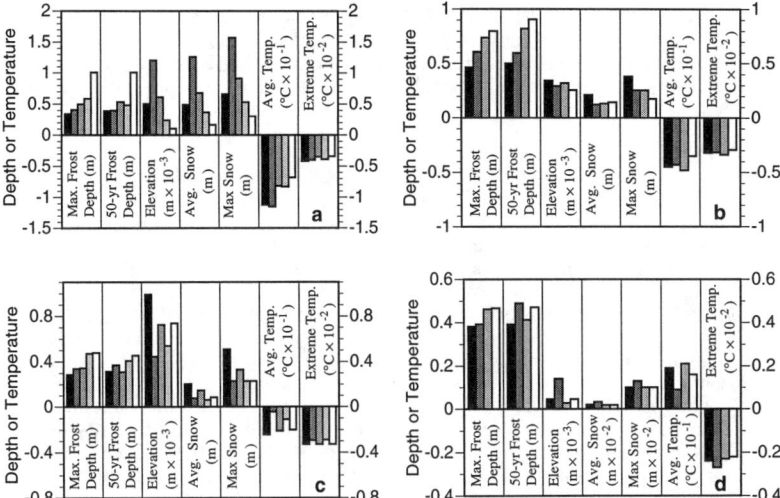

FIGURE 7. *Comparison of maximum depth under bare soil, 50-year return period frost depth, station elevation, average January–February snow depth, average maximum snow depth, mean January–February daily temperature, and average annual minimum temperature for stations (shown as separate bars) in a) New England, b) New York, c) West Virginia and d) Maryland.*

were recalculated based on snow-free conditions. Figure 8 summarizes these results for the New England and New York stations. Although the between-station variability in maximum frost depth within these regions is similar to that for the ambient snow conditions, the relationship between air temperature and soil freezing depth becomes pronounced. Clearly, the deepest frost penetration occurs at the coldest sites. While these findings support the use of empirical air temperature-based

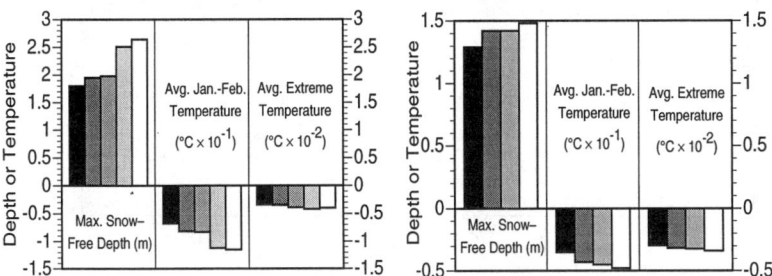

FIGURE 8. *Comparison of maximum frost depth under snow-free bare soil, mean January–February daily temperature, and average annual minimum temperature for stations in a) New England and b) New York.*

soil freezing models (e.g. Aldrich and Paynter, 1953) for snow-free conditions, they also illustrate the ability of snow cover to substantially moderate soil freezing depths. Therefore any spatial interpolation of the modeled soil freezing depths used to generate Figure 4, should consider small scale variations in the depth and duration of snow cover.

Sensitivity to Soil Characteristics

The frost depths shown in Figures 4, 7 and 8 depict results for soils having a clay content of 15% and a porosity of 45%. Non-clay soil particles are assumed to be quartz-based. It is further assumed that air occupies 10% of the available soil volume and that any remaining pore space is filled with water. Although this limits the model's application to wet soil conditions, this restriction is of little consequence over regions such as the northeastern United States, where high levels of soil moisture are typical during the period of soil freezing. In order to quantify the effect of differing clay contents and porosities on annual maximum frost depths, a geographically representative set of 30 stations was selected At each of these sites separate frost depths corresponding to the 2-, 5-, 10-, 25-, 50- and 100-year return interval were calculated using soil porosities ranging from 30 to 60% in increments of 5%. Similarly frost depths were computed for clay contents of ranging from 2 to 50%, holding porosity constant at 45%. Separate sensitivity analyses were conducted for bare soil, sod and snow-free bare soil.

Modification of the the clay content had little effect on the depth of soil freezing. In general, the difference in maximum soil freezing depth between the standard (15% clay content) and either clay content extreme (2 or 50%) was less than 5%. Changes in porosity, and thus water content, had a more pronounced effect on the maximum depth of frost penetration. Figure 9 shows these differences in maximum frost penetration as a ratio (multiplied by 100) of the annual maximum freezing depth based on the given porosity to that which occurred using the 45% standard porosity. Since the station-to-station and return period-to-return period

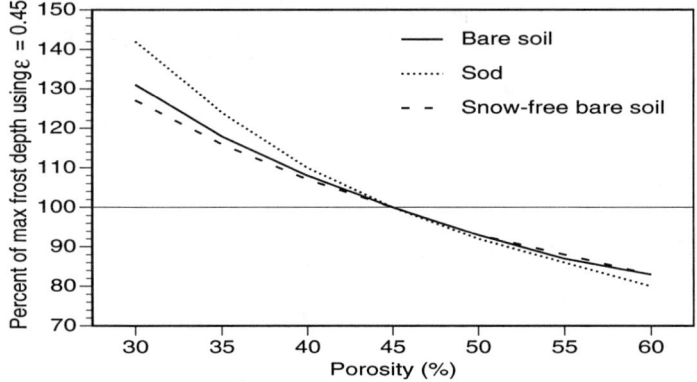

FIGURE 9. *Graph of adjustment factors (% of the maximum frost depth using a standard soil porosity of 0.45) used to convert the maximum frost depth values presented in Figure 4 to values representative of a site-specific soil porosity. Adjustment factors are medians over the 30 stations and 6 return periods.*

differences in these ratios was quite small (generally ≤ 0.03) the values shown in Figure 9 are the medians over the 30 stations and 6 return periods. In application, Figure 9 can be used to adjust the maximum frost depth values presented in Figure 4 to values representative of a site-specific soil porosity. For example, to compute a 100 year return period frost depth for a bare-soil site in New York City having a soil porosity of 55%, the 80 cm value given presented in Figure 4 is multiplied by 0.89 yielding a depth of 71 cm.

<u>Summary</u>

Measured soil freezing depths are not available at a network of stations that is sufficiently dense for most climatological applications. Therefore, a soil freezing model capable of estimating annual maximum soil freezing depths quite accurately using only those meteorological variables measured by the relatively dense cooperative observer network, was used to develop an extreme-value climatology of annual maximum frost depths in the northeastern United States.

Across the northeastern U.S., maximum annual frost depths under bare soil tend to be deepest in eastern Maine, where the estimated frost depths associated with the 50 and 100-year return periods exceed 110 cm and 120 cm, respectively. A secondary frost depth maximum occurs in north-central Pennsylvania, where the 50 and 100-year maximum estimated frost depth return periods exceed 90 cm and 100 cm, respectively. This pattern of deepest maximum frost depths is repeated for sod-covered surfaces. However, frost depths are generally 55 - 75 % shallower under sod.

South of the New York–Pennsylvania border, there is a general decrease in maximum frost depth with latitude. Maximum frost depths also decrease toward the coast. This pattern tends to reverse to the north, as maximum frost depths generally decrease as latitude increases. Except for Virginia, southern Maryland and Delaware, the shallowest maximum frost depths are found along the shores of the Great Lakes, over the Tug Hill Plateau region of New York (to the east of Lake Ontario) and in extreme northern New Hampshire. Exceptions to this general northward decrease in maximum frost depths occur in the Champlain Valley along the New York–Vermont border and throughout much of southern New Hampshire. Snow cover is the likely cause of this spatial pattern of maximum frost depth in the northern part of the domain.

Acknowledgment
The encouragement of Jay Crandell of the National Association of Home Builders Research Center was greatly appreciated. We also thank Megan McKay for her computer programming assistance. This work was supported by NOAA Grant NA16CP-0220-01.

<u>References</u>

Aldrich, H.P., and Paynter, H.M. (1953). "Analytical Studies of Freezing and Thawing of Soils." First interim report. U.S. Army Corps of Engineers, New England Division, Arctic Construction and Frost Effects Laboratory (ACFEL) Technical Report 42., 66 pp.

Benoit, G.R. and Mostaghimi, S. (1985). "Modeling soil frost depth under three tillage systems." *Trans. ASAE*, 28(5), 1499-1505.

Cary, J.W., Campbell G.S., and Papendick, R.I. (1978). "Is the soil frozen or not? an algorithm using weather records." *Water Resour. Res*, 14(6), 1117-1122.

DeGaetano, A.T., Eggleston, K.L., and Knapp, W.W. (1995). "A method to estimate missing daily maximum and minimum temperature observations." *J. Appl. Meteor.*, 34, 371-380.

DeGaetano, A.T., Wilks, D.S., and McKay, M. (1996). "A physically-based model of soil freezing in humid climates using air temperature and snow cover data." *J. Appl. Meteor.*, 35, 1009-1027.

DeGaetano, A.T., Wilks, D.S., and McKay, M. (1997). "Extreme-Value Statistics for Frost Penetration Depths in the Northeastern U.S.", *Journal of Geotechnical Engineering*, [in press].

Flerchinger, G.N. and Saxton, K.E. (1989). "Simultaneous heat and water model of a snow-residue-soil system - I. Theory and development." *Trans. ASAE*, 32(2), 565-571.

Guymon, G.L., Berg, R.L., and Hromadka, T.V. (1993). "Mathematical Model of Frost Heave and Thaw Settlement in Pavements." U.S. Army Corps of Engineers, Cold Regions Research & Engineering Laboratory (CCREL) Report 93-2, 126 pp.

Ricard, J.A., Tobiasson, W., and Greatorex, A. (1976). "The Field Assembled Frost Gage." U.S. Army Corps of Engineers Cold Regions Research and Engineering Laboratory Technical Note, 7 pp.

USDA, (1941). " Climate and Man, Yearbook of Agriculture 1941." U.S. Government Printing Office, Washington, DC, 1248 pp.

Wilks, D.S., (1995). "Statistical Methods in the Atmospheric Sciences." Academic Press, San Diego, CA, 464 pp.

SUBJECT INDEX

Page number refers to the first page of paper

AUTHOR INDEX

Page number refers to the first page of paper